TAKING THE GUESSWORK OUT OF AQUACULTURE

Farm Business Management

FBM20101NC

A how-to guide to planning a profitable fish farm.

JOHN DEERE

FREE CATALOG

TAKING THE GUESSWORK OUT OF AQUACULTURE

A how-to guide to planning a profitable fish farm

FARM BUSINESS MANAGEMENT

PUBLISHER

Farm Business Management Series (FBM) is a series of manuals created by Deere & Company. Each book in the series was conceived, researched, outlined, edited, and published by John Deere Publishing.

DEERE & COMPANY, JOHN DEERE PUBLISHING
TIAC Building, 1300 19th Street, East Moline, IL 61244
1-800-522-7448

JOHN DEERE PUBLISHING STAFF
John E. Kuhar
Lori J. Dhabalt
Cindy S. Calloway

AUTHORS: *Martin W. Brunson* holds a B.S. in biology and M.S. and Ph.D. in Aquaculture and Fisheries Management. He began his career as an assistant professor in Aquaculture and Fisheries at Louisiana State University, moving to Mississippi State University in 1988. He has conducted extensive research in aquaculture of catfish, crawfish and sunfish, as well as recreational fisheries management in small impoundments. He is currently Extension Leader and Professor in Wildlife and Fisheries at Mississippi State University, and works closely with the aquaculture industry nationwide in meeting their educational and technical assistance needs. He has authored a comprehensive set of over 150 fisheries and aquaculture extension publications, as well as 75 technical and scientific articles.

Bob Ratliff is an agricultural journalist with experience in radio, TV and print. From 1980 until 1990, he served as farm director of the Progressive Farmer Network, a regional radio network providing agricultural programming to stations in the Midsouth. His career also includes four years as editor of *MAFES Research Highlights*, a quarterly magazine published by the Mississippi Agricultural and Forestry Experiment Station. His articles about the work of MAFES researchers have appeared in *The Catfish Journal, Cotton Farming, Delta Farm Press* and other agricultural publications. He currently serves as Mississippi State University's science writer, providing print and broadcast coverage of the university's science-related teaching and research activities.

PHOTOGRAPHY: Photos by Martin W. Brunson unless otherwise noted.

COPY EDITOR: *John Harvey* is a veteran agricultural writer with a background in both farm magazines and ag public relations. He was an editor at *Successful Farming* and *Farm Journal*, was the writer for the USDA 1976 Bicentennial Yearbook of Agriculture, co-editor of the 75th anniversary book ("Farm Magazines, Milestones & Memories") of the American Agricultural Editors' Association, and wrote the book, "Classic Tractor Collectors: Preserving and Restoring Farm Power from the Past," published in 1994. His degree in Agricultural Journalism was earned at the University of Missouri-Columbia.

CONSULTING EDITOR: *William J. Varano*, M.S., C.P.Ag. is a certified consultant. His qualifications are in the fields that follow: Agricultural, agribusiness, management planning, design and engineering, sourcing, technical advice, project management and structuring. Agronomic: agronomic and environmental planning, nutrient management, groundwater, soil fertility and pesticides, economics of soil management. Engineering: complete facility design of hogs, poultry, beef, dairy, horse, aquaculture building design and construction management. Complete design of feed processing facilities, grain storage and handling, ventilation design, research facility design and custom fabrication of equipment and buildings. Recirculating aquacultural fish propagation and growth systems. Management: agribusiness planning, forecasts, capital and project management, construction, project structuring, technical advice, equipment sourcing, new technology integration to business planning.

SPECIAL ACKNOWLEDGEMENTS: The authors extend thanks to LaDon Swann, Purdue University, for contribution of the Species Profiles section and for his tireless work in providing leadership for AquaNIC. Thanks also are extended to graphic artist George Taylor, whose watercolor was used for the illustration on page one and to Jimmy Avery, Kevin Fitzsimmons and Gary Fornshell for photo contributions to AquaNIC which were used in this publication.

PUBLISHER: Farm Business Management (FBM) texts and visuals are published by John Deere Publishing, Dept. 374, TIAC Building, 1300 - 19th St., East Moline, IL 61244. This text and its supporting materials are intended only as educational media.

FOR MORE INFORMATION: This text is part of a complete series of texts and visuals on agricultural management entitled Farm Business Management (FBM). Instructor guides and student guides are available for many subjects. For information, request a free Catalog of Educational Materials. Send your request to: John Deere Publishing, Dept. 374, TIAC Building, 1300 - 19th St., East Moline, IL 61244 or call 1-800-522-7448.

ISBN 0-86691-252-5

CONTENTS

PREFACE:

During the past two decades, aquaculture has been the fastest growing segment of agriculture in this country and many other nations. Today, interest in aquaculture continues to grow. U.S. fish farms cultivate about 100 different species of fish and shellfish. The total on-farm value of products sold by the U.S. aquaculture industry each year is about $1 billion.

American aquaculture helps satisfy seafood demand, provides an alternative agricultural opportunity, provides jobs for displaced commercial fishermen and the rural community, and helps reduce the nation's trade deficit.

This manual provides an overview and snapshot look at the various U.S. aquaculture enterprises. It is intended to serve as a basic guide to prospective producers of aquaculture products in evaluating their potential for success in such an enterprise.

While it is not intended to be a day-to-day reference manual for anyone involved in the business of aquaculture, it can be used as a beginning point for making decisions about selecting an aquaculture enterprise that fits an individual's resources and capabilities.

A Brief History of Aquaculture

1

Aquaculture, the husbandry of aquatic animals, is a time-honored and long standing means of producing animal protein for human consumption. No one knows exactly when or where the first fish farmers stood back to admire the result of their efforts, but most aquaculturists generally agree that fish farming began in the Far East.

The first known written record of an aquaculture enterprise is an account of carp production in China in 475 B.C. Later references provide evidence of fish culture in China well before the time of Christ.

Ancient Chinese fish farmers
From a water color by George Taylor

From China and surrounding regions, aquatic animal farming spread around the world. Today, some type of aquaculture is practiced in almost every geographic region of the globe except those where the climate is too cold for open water.

In Europe, the Romans cultured oysters as long as 2,000 years ago, but the significant development of aquaculture on the European continent did not occur until the Middle Ages, when carp and trout production became well established. Since that time, aquaculture has been highly developed in Europe, especially in the Scandinavian countries, where salmon production is a major enterprise.

Aquaculture in Africa and Latin America has been minimal compared to other regions, but small-scale family-based operations are common. In those tropical and semitropical regions tilapia, a warm-water food fish which is becoming the predominant species in recirculating water systems in the U. S., is the dominant species. Recently, however, Central and South American producers have had significant success with shrimp.

North America has become the most recent and most productive and profitable region for aquaculture industries. The early focus in North America was on oysters and trout. Oyster farming was a remnant of European oyster culture, but never really became a viable production industry. Trout production, on the other hand, grew in scope and technology. U.S. producers quickly became leaders in trout hatchery and production techniques. The early emphasis was on raising trout species for human consumption, but the need soon arose to produce trout for stocking streams and native waters. Thus, early aquaculture in the U.S. focused on hatching and propagating trout and other species.

By 1870, 19 of 37 states, plus the territories of Colorado and Kansas had some kind of aquaculture enterprise. Many state fish commissions set up public hatcheries to produce fish for stocking native waters.

In 1871, the Federal government established the U.S. Commission on Fish and Fisheries, and subsequently produced millions of fish for release into natural waters. About the same time, the American Fish Culturists Association was established, and ultimately grew into what is today known as the American Fisheries Society, the widely acknowledged professional organization for fisheries and aquaculture professionals in the U.S.

Oysters have been cultured at least 2,000 years
Photo by Jimmy Avery

After aquaculture and fisheries development in the U.S. was set into motion, investigations into the culture of a variety of species began in earnest. The first Federal appropriations for fish research were made to a laboratory at Woods Hole, Massachussetts.

In the 1920s and 1930s, commercial aquaculture established a permanent foothold in the American economy and society, beginning with bait minnow culture, and expanding to include several varieties of sport and food fish.

In the 1930s, President Franklin D. Roosevelt established a "Farm Pond" program. This program provided government assistance to farmers in the construction and stocking of ponds with fish to increase their income.

Although established, American aquaculture struggled for survival for another two decades, really only thriving in the forms of state or federal hatchery programs.

U.S. aquaculture truly came into its own with the early success of the catfish industry during the 1960s and 1970s. Since that time, many species have been cultivated with varying degrees of success. Although catfish remains the most successful aquaculture enterprise, crawfish, rainbow trout, bait fish, striped bass and its hybrids, several types of sport fish (including largemouth bass, bluegills, crappie and yellow perch), tropical aquarium fish and tilapia, are all established as viable aquaculture industries. The annual on-farm value of aquaculture enterprises in the U.S. totals about $1 billion.

General Culture Approaches

2

Most fish farms concentrate on growing only one life stage of the species involved. Others may culture several, or all, life stages. Some farms specialize in producing "seed stock," juveniles that are sold to others that specialize in growing the seed stock from the juvenile stage to market or "food size" animals. Food size animals are sold to either processors for preparation and movement into the food distribution system, or sold directly to consumers or restaurants.

For most species, seed stock production involves some method of breeding, or spawning, adults to produce eggs. Spawning techniques can be as simple as placing pairs of mature adults into ponds or tanks and allowing them to spawn naturally, or as complex as manually extruding the eggs and sperm from brood stock and fertilizing them by hand in laboratory-like settings.

If brood stock are allowed to spawn on their own, the eggs are usually removed from the spawning pond or containers immediately after fertilization and placed in controlled environments called "hatcheries." In the hatcheries, the eggs mature and hatch into the early life stage of the cultured species. Hatcheries take many forms, ranging from large complex buildings with elaborate water delivery and filtration systems to small open ponds where water and other environmental conditions are carefully monitored.

Fish eggs in hatchery

One thing all hatcheries have in common is significant or even complete environmental control. The eggs and juveniles are never exposed to any unwanted external predators or foreign materials that may harm them or pose a hazard to humans consuming them at a later date.

The next step in the production process is movement from a hatchery to a "grow-out" facility, where the fish are grown from juveniles to adults. Growout facilities may range from small earthen or concrete raceways for trout, in-ground or above-ground circular or rectangular tanks for tilapia, to 12-15 acre earthen ponds for catfish.

Juveniles are stocked into production systems at various densities, depending on the species, and are usually fed a high quality, nutritionally complete manufactured feed until they reach a size demanded by the respective markets. For most species, the time required to grow from juvenile size to market size is between 6 and 24 months. This is an intensive management period for the fish farmer, since fish are typically stocked at high densities. Constant vigilance is required, and close daily attention must be paid to water quality requirements, such as dissolved oxygen levels and ammonia build up.

Since the animals are confined to the water in which they are grown, maintenance of the quality of that environment is essential for farming success.

It is this basic requirement of aquatic animals, the need for a "clean" and conducive growth environment, that results in one of the great attributes shared by all farm-raised aquatic products: FOOD SAFETY.

Because aquatic animals are intolerant of pollutants and other contaminants in water, aquaculturists must take great care to preserve the integrity of the water in which they produce their crop. The result is a product that not only tastes good and offers a healthy dining alternative, but one that also is clean, safe to eat, and free of pollutants.

In an effort to maintain and preserve the love affair between American consumers and aquaculture products, most of the major aquaculture commodity producer associations have adopted some type of Quality Assurance (QA) program and fish farmers have incorporated QA principles into their daily production practices. These QA principles are based on recently developed food safety assurance standards and procedures established and required by the U.S. Food and Drug Administration (FDA) and the U.S. Department of Agriculture (USDA).

Critical considerations in any aquaculture QA program include:

- **Selecting a production site**
- **Selecting a water supply**
- **Managing water quality during production**
- **Managing hatchery and brood stock**
- **Feeding practices and nutrition**
- **Managing fish health**
- **Harvesting and transporting of fish**
- **Processing of fish**

Proper care and procedures at each of these critical consideration points decrease the possibility of environmental contamination or drug and chemical residues in fish.

Some other meat animal industries adopted such QA procedures only after experiencing problems with failed consumer confidence, but the aquaculture industry, which has not had a consumer confidence problem nor a food-safety crisis, has proactively embraced the concept of Quality Assurance. The industry recognizes the vital and fragile nature of the consumer-producer relationship. Consequently, the aquaculture industry does not anticipate any kind of drop in consumer confidence, and it works hard from the farm to the processor to the retailer and distributors to avoid the possibility of a failed relationship with its consumers.

Hybrid striped bass

Once the animals reach market size, usually 1-2 pounds for most fish species, or 15-40 per pound for shrimp and crawfish, they are harvested from the production facility. Nets are usually used to remove the animals from the growout ponds, tanks or raceways. A typical harvest net for catfish or hybrid striped bass, called a seine, may be as long as 1 mile, while harvest of tilapia or trout from raceways and tanks requires only hand held dipnets or baskets.

For catfish, tilapia, hybrid striped bass, and other fin fish species grown in ponds, the fish are corralled into one corner of the pond with the seine and large baskets suspended from truck-mounted booms are used to remove the fish from the pond. They are then loaded onto transport trucks for live-hauling to processing plants.

Harvest of crawfish is unique, employing baited wire mesh traps similar to the crab and lobster traps used in coastal areas by recreational and commercial fishermen. Live crawfish are placed in plastic mesh vegetable sacks and transported immediately to either processing plants or live-sale markets. Crawfish are one of the few aquacultured species that can survive out of the water!

Regardless of harvest method, most animals are immediately transported live to the processing facility where they are prepared for storage and subsequent shipment through the distribution network to retailers and consumers. Species such as bait minnows are delivered live to consumer markets.

Some species are slaughtered and immediately processed into various forms such as fillets, steaks or whole-dressed. They are then individually quick frozen (IQF) or shipped ice-packed whole for fresh markets. Others, such as hybrid striped bass and most tilapia, are immediately chilled in ice baths and shipped whole, either on ice or quick frozen. Regardless of the species, the time between harvest and freezing or ice-packing is usually just a few hours, guaranteeing that only the freshest possible product enters the marketplace.

Like the farmers, processors have embraced QA practices, and submit their processing procedures and their products to very stringent quality control standards. Processors work hard to ensure that the wholesome, high quality, clean and safe products they receive from the farmers remains wholesome, high quality, clean and safe until it reaches the dinner plate in a restaurant or their consumers' homes.

In December 1997, the FDA put into effect Hazard Analysis of Critical Control Points (HACCP) to insure the uniform cleanliness and purity of processed seafood products. This applies to processors, warehouses, importers, and all handlers of processed seafood.

This relatively new kid on the American farm scene has inherited a major challenge as it strives to fill the void created when demand began to exceed naturally available fish and shellfish products. But aquaculture, with its dedicated production, processing and marketing infrastructure has proven itself more than up to the challenge, and the consumer love affair with fish and shellfish products is not only alive and well, but grows more intense every day.

A DAY IN THE LIFE OF A FISH FARMER

For most fish farmers, day and night often run together, since managing a fish farm is a 24 hr job during many parts of the year. The following is an example of a day in the life of a hypothetical fish farmer.

5:00 a.m.The alarm rings early on the fish farm. Early morning tasks include the daily checking of oxygen levels. Preparation of production units for harvest is another early-morning task at certain times of the year.

6:00 a.m. Dawn sees the fish farmer on the ponds or production units checking dissolved oxygen (DO) levels. This is the most critical time of the day for DO levels. Others on the farm may also be engaged in seining or harvesting of fish to be sent to markets or processing plants. During warm months, early morning harvest and transport are often preferred.

7:00 a.m. Predacious birds such as herons and cormorants arrive from their nighttime roosts, ready to begin feeding on the fish in the production units. Action must be taken to deter the feeding activity, including scaring the birds into leaving the premises.

9:00 a.m. The daily task of feeding fish usually begins about mid-morning. On larger farms, feeding can require several hours of effort, while feeding on smaller farms or raceways and tank systems may be done in a relatively short period of time. Regardless of the time involved, feeding is perhaps the most critical task on the farm.

10:00 a.m. Contacts must be made with market outlets to ensure continued sales of fish and products. Time spent on the telephone with buyers and other market outlets is a wise investment in the future.

10:30 a.m. Right in the middle of the phone conversation with the processing plant, the farmer gets word that the feed truck has broken down! The conversation is cut short as the quick response to failed equipment begins. Equipment maintenance and repair are necessary parts of managing any fish farm, and quick turnaround avoids costly delays.

12:00 p.m. Hopefully, if all equipment is functioning and morning management chores are completed, the producer is able to take a break and eat lunch. This is sometimes a sandwich back at the office or at home, but more often simply a soft drink and crackers (sardines, too with luck!) in the truck moving from place to place on the farm.

1:00 p.m. If a lunch break was taken, it's time to get back to work! On many farms, feeding and the eternal equipment maintenance task is resumed. If there have been no breakdowns today, there will at least be routine and preventive maintenance such as checking of pumps, bearings, aerators, DO meters, nets, seines, etc.

3:00 p.m. Samples of fish from ponds or production units scheduled for harvest must be taken and delivered to the processor for flavor sampling. This is a critical marketing consideration, since catfish processors, for example, will not purchase fish unless they have passed flavor tests administered by the processors themselves.

5:00 p.m. Late afternoon preparations for the night's vigil must be made. A quick check and calibration of DO meters, as well as fueling the truck, checking aerators, electrical connections, etc. occupy full attention.

7:00 p.m. DO readings are taken at dusk and then periodically throughout the night.

8:00 p.m. With any luck, it's dinner time, and the opportunity to relax with family is finally here.

9:00 p.m. After the kids are in bed, it's time to update daily records on the computer, review financial transactions and plan the day ahead. This is also the time to review trade magazines and catch up on the latest extension publications that provide results and practical applications of research projects. A successful fish farmer knows that learning is a lifelong process, and takes advantage of any opportunity to gain new knowledge.

10:00 p.m. After a final check on the farm, provided that all is well, it's time for bed. But if it's summer, the sleep will be short, because midnight and 2 am DO checks are the rule of thumb to be sure that oxygen depletions don't occur!

5:00 a.m. The alarm rings early ....

3 Types of Aquaculture Enterprises

In general, aquaculture production can be classified as either "intensive" or "extensive". As the name implies, intensive culture involves raising the product in densities that are higher than could be supported by Mother Nature, and requires significant management input. Extensive culture involves raising the product in low densities, often on naturally occurring foods, rather than manufactured feeds. Most aquaculture enterprises are intensive in nature.

Aquaculture takes place in a variety of facilities. The most common commercial production unit is the pond, but other types of rearing units are common, especially for specialty fish, research or captive breeding projects, and small-scale operations.

Earthen Ponds

Ponds sculpted from the earth are grouped into three categories:

Construction of a Levee pond

1. **Embankment pond** - a pond created by constructing an embankment or dam in a naturally occurring basin that collects rainfall runoff from the watershed.

Ariel view of catfish ponds

2. **Excavated pond** - a pond created by digging a hole, usually on flat or gently rolling terrain, but sometimes in land with high slopes. The dirt removed from the hole is used to build exterior levees if needed. The bottom of an excavated pond is usually below the level of surrounding terrain.

3. **Levee pond** - a pond built on level land by pushing dirt from the center of the pond area to form exterior levees, usually on all four sides. The bottom of a levee pond is usually at or near the level of the surrounding terrain.

Raceways

Raceways used for tilapia production

Raceways are narrow, rectangular units that allow a frequent water exchange by allowing water to flow through the units, usually by force of gravity, but sometimes through a pump system. The ratio of depth:width:length is often 1:3:30 (for example, a raceway 3 feet deep might be 9 feet wide and 90 feet long). Raceways are frequently used for rearing trout, and lend themselves to use in areas with steep slopes, but can be used in any topography.

Two basic raceway types are used:

1. **Concrete (or other material)** - can be constructed in almost any soil type and terrain. These are typically arranged in a parallel series, and many units can be placed side by side in a small area of land.

2. **Earthen** - usually excavated from soils that will hold water. More land is required to construct earthen raceways than for concrete or other materials, since the widths of the dividing walls, called levees, must be greater for earth than for concrete.

Indoor tank system

Tanks and Vats

The possible combinations of shape, size and material are endless when considering the use of tanks or vats to grow fish. Essentially any container that will hold water can be used, but years of experience have proven that circular or rectangular shaped units are best. In facilities where tank or vat culture is practiced, it is very important that the waste products from the fish be removed to maintain adequate water quality. This is done in one of two ways, either by a constant or periodic flow of fresh water into the system, flushing the waste products (flow-through systems) or by a sophisticated filter system that allows water to be continually cleaned and reused (recirculating system).

Although this type of system lends itself well to situations where available land is minimal, or where pond or raceway construction is not possible, management is usually very intensive and the cost of production for most species is increased. Such facilities are used more frequently to HOLD fish, rather than to actually GROW fish. Exceptions include hatcheries and research facilities, as well as a few small-scale specialty product operations. In these cases, the increased cost of culture in contained systems is justified to obtain desired results. In research situations, the primary concern is generation of research data, not necessarily profitable production. In hatcheries and some small-scale specialty industries, such a system is vital to survival of the fish and success of the operation.

Some new large-scale warm-water tank systems are being developed that make production of tilapia and striped bass more economically feasible by increasing the density of fish per gallon of water.

Cages

Cages are floating or suspended enclosures made of some type of net-mesh in which fish are grown or held. These are usually floating in another body of water such as a pond, lake or bay. Shapes include round, rectangular and square. They

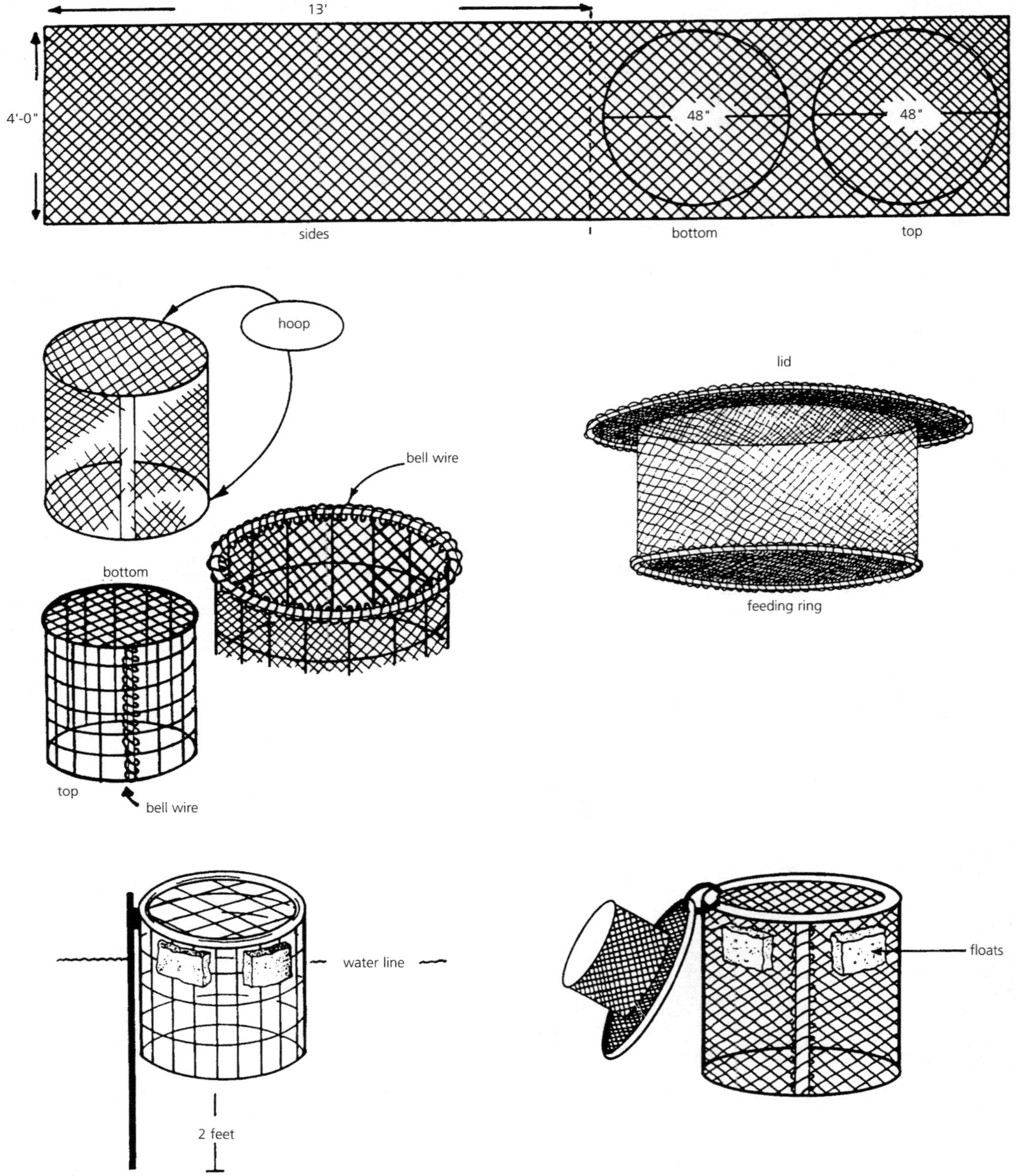

Typical Cages
Courtesy of Southern Regional Aquaculture Center

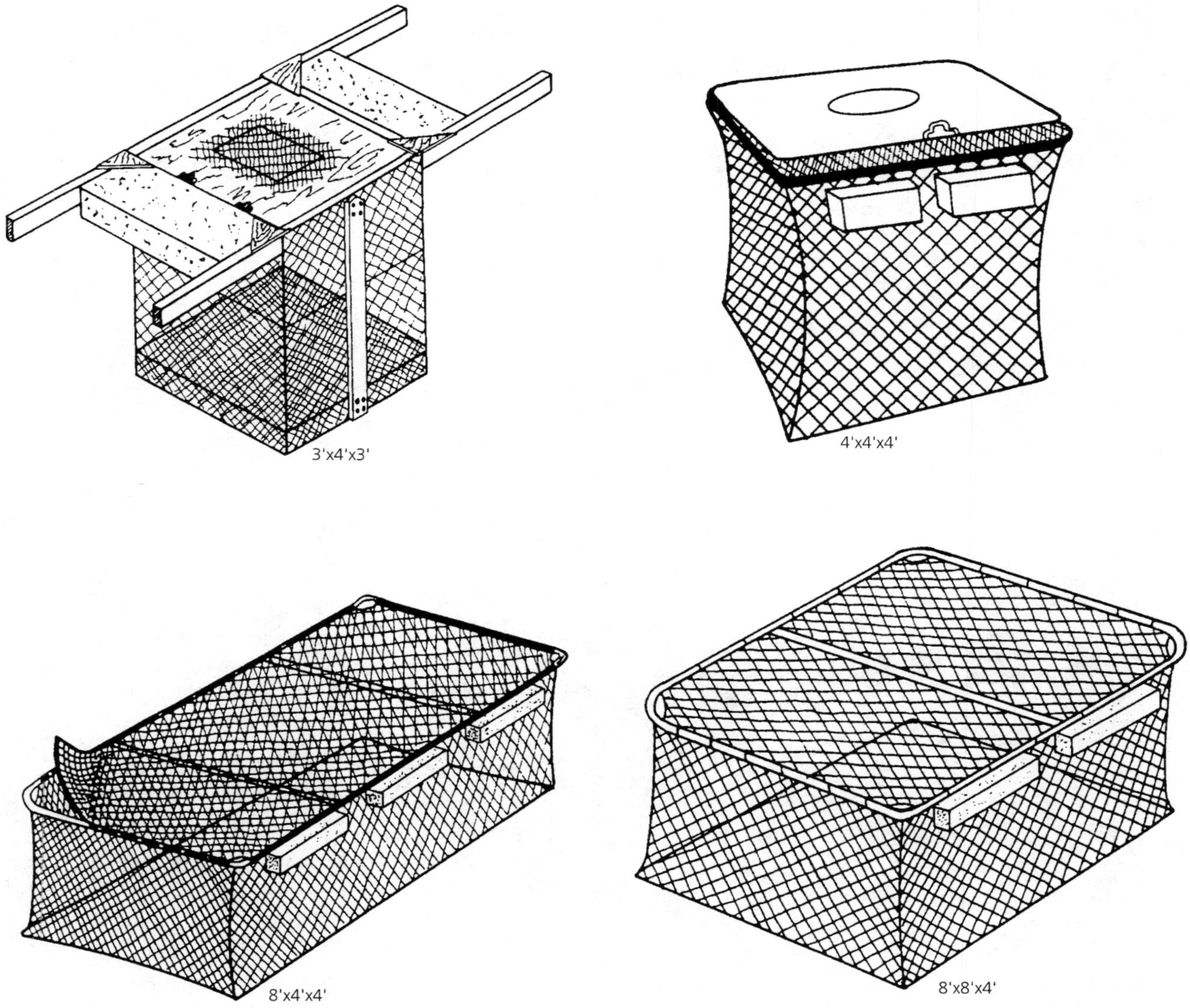

Other common cage designs
Courtesy of Southern Regional Aquaculture Center

are made of durable materials such as vinyl or vinyl covered wire, plastic or wood. A typical cage is 4 feet deep, 4 feet wide, and 8 feet long, but sizes vary and large cages called pens are often used to culture salmon.

Cage culture lends itself to small-scale operations with pre-existing bodies of water that would not otherwise be suitable for fish culture. Several species are successfully cage cultured, but few large scale operations exist in the U.S. except for the net-pen salmon facilities of the Atlantic and Pacific coasts.

4 Commercial Catfish Production

Channel catfish

Although more than 100 species of aquatic animals are grown commercially in the U.S., only a few represent highly significant economic importance. The types of culture systems employed by fish farmers to produce these crops of fish and shellfish are highly varied, and it is beyond the scope of this text to detail all the systems and approaches. As an example, however, it will be helpful to look closely at the largest and most valuable fish species — the channel catfish industry.

Catfish production is far and away the most prominent and successful component of the US aquaculture scene, accounting annually for over one-half of the total value of US aquaculture. Catfish production is centered in the four southern states of Alabama, Arkansas, Louisiana and Mississippi, and encompasses over 160,000 acres of water nationally. In 1997, production and sales topped 500 million pounds, with processor sales exceeding $500 million nationally, both all-time records.

Mississippi leads the nation with about 60 percent of the total U.S. catfish acreage, followed distantly by Arkansas, Alabama and Louisiana. Two Mississippi counties, Humphreys and Sunflower, each contain more acres of catfish ponds than any other single state in the nation! Mississippi produces about 72 percent of the catfish produced nationally and processes roughly 75 percent of the fish. The state does this with about 18 percent of the nation's catfish farming operations.

Catfish production in Mississippi is centered in the fertile Delta area, which contains 96 percent of the state's acreage. The Delta system of catfish farming is characterized by 15- to 17-acre ponds having levees on all four sides, by water being supplied by large wells, and by farms having several hundred acres in fish production.

The near-level land and abundant groundwater found in the Delta give the area a natural advantage in the production of catfish. Although small in comparison, non-Delta production has also increased. Some areas with relatively level land have put in Delta-type farms, but most of these areas have watershed ponds and depend on rainfall for fresh water. Production per acre is less on watershed farms, and marketing opportunities in small producing areas include more pond-bank sales and more farm processing to add value to the fish.

The future of the farm-raised catfish industry is virtually unlimited. Market development is one of the major marketing obstacles to overcome. Major objectives for the industry include having farm-raised catfish available in the form, at the place, at the time, and at prices consumers are willing and able to pay.

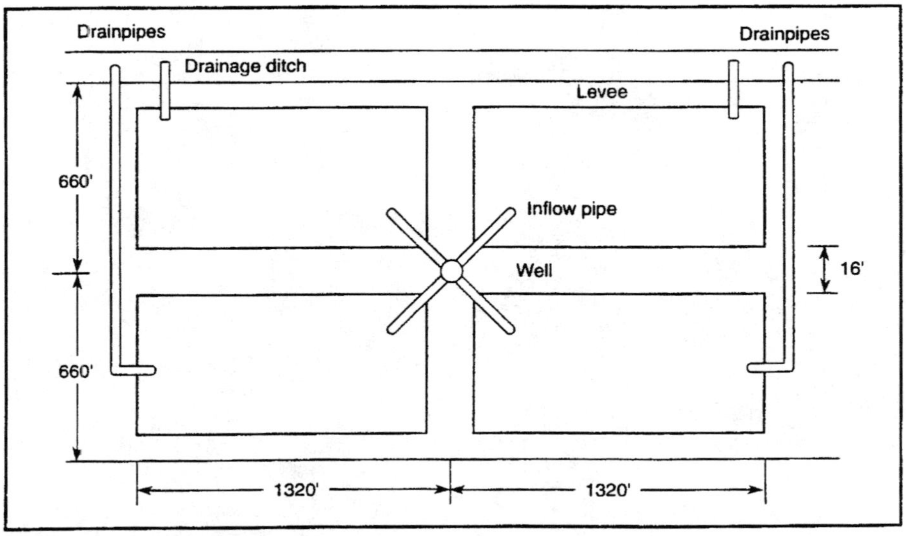

Catfish pond layout

Production Process

The production process begins with careful selection and mating of quality brood stock. Once the eggs are laid and fertilized, they are collected and placed in controlled hatchery tanks where they are closely monitored. The eggs hatch in 7 days at a controlled temperature of 78°F. When the eggs first hatch, the young fish are called "sac fry" because their yolk sacs are attached to their abdomens.

The sac fry live off food stored in the yolk sacs. As the yolk sacs are depleted, the fish begin to swim and are ready to be put in a pond where they grow into fingerlings. When the fingerlings are about 4 to 6 inches long, they are transferred to catfish ponds at various rates per surface acre of water.

Feeding

Catfish are fed a high protein, floating feed ration produced by several catfish feed mills located in the producing regions. This protein diet consists of soybean meal, corn, wheat and fish meal. Annual feed requirements are estimated to be about 5 tons of feed per surface acre of water. Catfish convert feed at an average yield of about 1 pound of fish for every 2 pounds of feed. Compared to other feed-converting animals, catfish produce one of the highest yields per pound of feed.

Water Quality

The oxygen content of the water must be checked on a regular basis since lack of oxygen can kill fish in a matter of minutes. When the oxygen content begins to reach a dangerously low level, such as on hot summer days, producers use paddle wheels or other devices to aerate catfish ponds. This technique introduces air into the water and replaces the oxygen that has been depleted.

Feeding fish with a blower feeder

Harvest and Marketing

Catfish harvest

By the time farm-raised catfish are 18 months old, they are usually ready for harvest. Averaging 1 to 1.5 pounds live weight, they are removed from the ponds by seines and placed in aerated tank trucks for live shipment to the processing plant. Some of the fish go directly to fish-out ponds or other fish markets, but the majority are processed. Catfish are kept alive right up to the time they are processed. Within just a matter of minutes they are processed and then placed on ice or frozen to temperatures of 40 degrees below zero, using an individual quick-freeze (IQF) method that allows the taste and quality to remain in the fish longer.

Processing plants have undergone a tremendous change in recent years. Today, 14 plants process 6 to 9 million pounds of catfish per week, with the capacity to produce more than 9 million pounds. In 1993 over 450 million pounds of farm-raised catfish were used for patties, smoked catfish, breaded catfish, and new precooked frozen dinners or entrees prepared as heat-and-serve items. Farm-raised catfish are taste-tested at the farm before the fish are harvested and again at the processing plant before the fish are unloaded. These quality control procedures ensure that the consumer receives a product of superior flavor.

Farm-raised catfish are marketed through three major channels: retail grocery store outlets, food service distributors and catfish specialty restaurants. Each channel accounts for about one-third of total catfish sales.

Consumer Demand

The per capita consumption of all fish and seafood in the U.S. in 1997 was 15 pounds, with the consumption of catfish exactly one pound, up from .41 pound in 1985. This represents only about 7 percent of the total fish and seafood consumed in the U.S., so the potential for growth is enormous.

Farm raised catfish has the qualities consumers demand

Farm-raised catfish has many nutritional advantages, including being low in fat, cholesterol and calories, and high in protein. In addition, farm-raised catfish is a good source of many vitamins and minerals.

Another advantage is that farm-raised catfish has less of the smelly fish odor associated with cooking fish indoors.

Farm-raised catfish have only large bones that are easy to locate and remove in whole fish. Catfish steak has only one bone section, which is easily removed after cooking. Catfish fillet strips and nuggets have no bones.

Farm-raised catfish is versatile and can be prepared in many ways, including fried, baked, broiled, sautéed, smoked, stuffed, served as hors d'oeuvres, in sandwiches, gumbos, paté, soups and quiches, sushi, Mexican style, or barbecued. The products are available year-round at local retail outlets. Taste tests have shown that most people who have tasted catfish for the first time rate it high on fish preference charts.

Economic Value

The farm-raised-catfish industry has given a big boost to the Mississippi economy with producers, processors, feed mills, suppliers and some retailers all located within the state. Catfish ranked fourth in 1997 among agricultural and forestry enterprises in the state, with an estimated total income of $382 million. The catfish industry in Mississippi and its adjoining states provides jobs for more than 12,000 people, with an economic impact of over $3 billion.

Quality Assurance

Farm-raised catfish is not exposed to external environmental factors, so producers can more easily control the quality of their product. Even in cultured products, however, potential contamination by pesticides, drugs, or other environmental factors is becoming a major consumer concern.

Catfish producers responded to increased consumer awareness and sensitivity with a Catfish Quality Assurance (CQA) program. This is an opportunity to demonstrate that catfish production practices are safe, carefully monitored, and that the safety and quality of farm-raised catfish can be assured. Catfish producers recognize that a producer-client relationship built on mutual trust and confidence is necessary for the survival of the industry.

Catfish producers are aware, however, that quality assurance goes far beyond the pond bank. Quality assurance begins, and hinges upon, activities at the production level. The quality of the product as it leaves the farm greatly influences the quality of the product reaching the consumer. Safety and quality cannot be "added" or increased at the processing or retail level; it must be guaranteed by the producers.

The strong and positive relationship that catfish enjoys with consumers must be maintained, nurtured and enhanced. CQA will serve to cement that relationship.

Producing catfish, like all other aquaculture commodities, is a highly specialized business requiring large capital outlays and unique production skills. If you are interested in producing any aquaculture crop, closely examine your skills and experiences, visit other farms, and use all the information and workshops available through the Cooperative Extension Service of your Land Grant university before you make an investment decision.

5 Required Resources and Investment

The decision to start an aquaculture operation is a major decision that will require many resources and a tremendous commitment of time and energy. The time to consider these requirements is BEFORE making the physical and financial commitment to establish a fish farm. In many cases, once ponds or other facilities are constructed on the land, it is extremely expensive to convert the property back to its former features.

Following are the most common resources needed to start a fish farm. Use this as a starting point in developing your own master list as an aid in your decision-making process.

Land

Regardless of the type of facility that you plan to establish, land is required. The type of land will depend upon the kind of production facility to be built. For example, a system of ponds will require a specific type of soil and topography, while a raceway or tank system can often be placed on land that is not suitable for ponds.

Enough land of suitable quality must be available to meet present needs as well as future expansion if your operation is successful. Consideration of neighboring resources and land uses is also important. For example, if the facility is located in an area of industrial activity, carefully examine the types of activities, potential pollutants, or other possible conflicts or compromises that may impact either your production or the quality, health or food safety of the fish you produce.

Facility and Pond Construction

If you choose to grow fish in ponds, careful consideration must be given to the location and type of land that will be used. The three primary areas to consider are:

1. Source, supply and quality of water

2. Soil type (must hold water)

3. Topography and proximity to other resources.

Pond construction

If ponds are not your choice of production facilities, many more options are available regarding location. The source and supply of water will still be important, but soil types and topography are less critical, since many tank and raceway systems can be placed on a variety of soil and land types.

If you decide to grow your crop in tanks, cages, or other production units besides ponds, you will need to add the costs and availability of either purchasing these units or fabricating them yourself.

Water Supply

Water will be required to successfully grow an aquatic commodity. There are three general sources of water, and the costs associated with obtaining water from each of these sources may vary widely. Specific considerations on water supply are provided in the chapter on "Site Selection."

Feeders and Feed Storage

Typical feed storage bin and feeder truck used in fish farming

Regardless of the type of system you establish, fish must be fed on a regular schedule. Feeding can be done by hand or mechanically. Use of mechanical devices such as blower feeders, automated or demand feeders, or other feed delivery machines will require purchase and maintenance of the selected equipment. Unless your operation is extremely small, some type of on-farm feed storage will be required. This storage must be dry and rodent free. Most operations choose to purchase feed in bulk, since such purchases usually provide a price discount. Depending upon the size of the operation and species cultured, feed requirements can range from a few hundred pounds to several tons of feed per day.

Fingerlings or Seed Stock

A source of seed stock will be required, and may be either on-farm production (with its own associated costs) or purchased from another enterprise. In either case, the costs of obtaining seed stock for growout to market size will be a major economic factor.

Feed

For many operations, feed represents over 50% of the total annual operating costs of production. Most species that are grown in the U.S. require daily feeding of a manufactured ration. Feed costs vary depending upon prices of ingredients. Most feeds used by U.S. producers are manufactured by commercial feed companies, rather than milled on the farm as is done with some other meat animal production.

Disease and Weed Control Equipment

Equipment for the application of herbicides, parasiticides and other therapeutics will be required, and may range from small sprayers to large application boats. In many production systems, especially intensive, recirculating systems or tanks culture, glassware, measuring devices and other laboratory type equipment may be needed.

Many operations set up small laboratories where water quality and fish disease analysis can be performed. A small microscope and a water testing kit with glassware may be required.

Oxygen Testing Equipment

Oxygen meters and test kits are among the most useful, often used, and necessary equipment items that a fish farmer can have. The chapter on "Water Quality" discusses the need for these items in detail.

Tractors, Trucks or other Farm Vehicles

In pond systems, tractors are absolutely necessary to move and power portable aerators, to mow and maintain levees, and to perform other tasks around the facility. Regardless of the type of operation, some type of vehicle and/or small power equipment will be required to perform day-to-day tasks. A pickup truck is almost indispensable, and other modes of transportation or equipment transport may also be necessary. Four-wheel ATVs have become integral, utilitarian components of most fish farms.

Aerators

Aerators and other oxygenation equipment will be absolutely necessary, regardless of the production system chosen.

Miscellaneous Equipment

Depending upon the production system of choice, many other items - large equipment to small supplies - may be required. Be sure to include these miscellaneous items in your cost calculations.

Feeds can be purchased in bags or in bulk
Photo by Jimmy Avery

THE INVESTMENT

The amount of investment required for fish farming varies and depends on many factors. Following is a list of costs common to most fish farms. Determine the approximate costs for your situation:

Capital Costs

	Amount
1. Land	________
2. Pond construction (about 6.2 cubic yd. per linear foot of levee for a 16-foot top), gravel, vegetation	________
3. Drains (pipe, valve, fittings)	________
4. Water supply	________
5. Building (production and/or storage and service)	________
6. Aeration equipment (paddle wheel, relift pumps, blowers, liquid oxygen, etc.)	________
7. Boat and motor	________
8. Trucks	________
9. Fish hauling tanks and agitators	________
10.Feeding equipment, feed storage bins	________
11.Tractors	________
12.Mowers	________

Capital Costs

	Amount
13. Oxygen testing equipment	________
14. Other water quality testing equipment	________
15. Seines, storage reels, crane	________
16. Dip nets	________
17. Waders and boots	________
18. Miscellaneous equipment	________
19. Taxes and insurance	________
20. Interest	________
21. Water filtration equipment for waste removal	________
22. Filtration equipment for biological system balance	________

Operating Costs

1. Feed	________
2. Fingerlings	________
3. Electricity	________
4. Fuel (diesel and gasoline)	________
5. Maintenance and repairs	________
6. Harvest labor	________
7. Transportation	________
8. Daily labor	________
9. Chemicals and drugs	________
10. Telephone	________
11. Miscellaneous	________
12. Interest on operating capital	________

The Decision Checklist

6

Fish farming is much more than just stocking fish in a pond, feeding and then reaping the profits. It is a very expensive and risky form of agriculture. Intensive fish culture requires almost 24-hour a day management. Unless you are willing to provide this type of management, you should look at another enterprise. To help you determine if intensive fish farming is feasible for you, a checklist is provided below. It does not cover all of the possibilities, but it does list most of the important considerations. Answering "yes" to all or most of the questions does not guarantee success, just as answering "no" does not mean failure.

Management	**Yes**	**No**
Do you already have suitable ponds or a pond site located?	___	___
Do you have most of the equipment (tractors, pumps, storage building, etc.) needed?	___	___
Do you have the necessary financial resources (about $3000/acre investment and $2500/acre annual production cost) needed?	___	___
Have you made an estimate of investment costs and annual cost and return?	___	___
Have you estimated the impact of changes in fish prices and feed costs on projected income?	___	___
Will current interest rates and interests costs on investment and operating capital permit a reasonable profit?	___	___
Will the expected profit provide an adequate return for your labor, management and risk?	___	___

Management	**Yes**	**No**
Is fish farming the best alternative for the land you intend to use?	___	___
Can you afford to forgo income until you sell your first crop (usually 15-24 months after starting)?	___	___
Have you looked at record systems available and picked one best suited to your situation?	___	___
Can you afford to absorb occasional losses?	___	___
Are you willing to devote the time and effort required?	___	___

Marketing		
Do you know of an established market for your fish?	___	___
Is there a market for your fish at the time of year you plan to sell them?	___	___
Will you have harvesting and transport equipment, or do you have a suitable arrangement for harvesting your fish?	___	___
Will you be able to harvest fish year round?	___	___
Do you have an alternative marketing strategy?	___	___

Physical Factors	Yes	No
Will the soil hold water?	____	____
Is the topography of the land suitable for pond construction?	____	____
Is there adequate ground water close enough to the surface for fish farming?	____	____
Is the water quality suitable for fish farming?	____	____
Is the pond area protected from flooding?	____	____
Are the drains in existing ponds adequate for rapid draining?	____	____
Can you prevent wild fish from entering the pond?	____	____
Is there daily access to the ponds, regardless of weather, for feeding, treating and harvesting?	____	____
Is the pond bottom suitable for harvesting (smooth and stump-free)?	____	____
Will someone live close enough to the pond to allow frequent observation and necessary management?	____	____

Production		
Are good quality feeds available at competitive prices?	____	____
Is there a source of supply for drugs and chemicals?	____	____
Are fingerlings available at competitive prices?	____	____
Can you make or purchase needed aeration equipment?	____	____
Is dependable labor available?	____	____
Is dependable diagnostic service available?	____	____
Do you have adequate storage facilities for feed?	____	____
Are you aware of the government agencies that can provide you with educational and technical assistance?	____	____

Risks	Yes	No
Are you prepared to handle problems:		
• Poor water quality?	____	____
• Diseases?	____	____
• Pesticide contamination?	____	____
• Poachers and vandals?	____	____
• Low fish prices and high feed costs?	____	____
• Personal stress resulting from financial loss?	____	____

Site Selection

7

Source, Supply and Quality of Water

There are two basic types of water sources: ground water and surface water. Ground water is supplied by wells drilled to underwater aquifers at varying depths, while surface water can be obtained by pumping from ponds, lakes and streams, or by catching rainfall runoff directly.

Ground Water

Pumping well water into fish pond

In most cases, from a quality standpoint, the preferred source is ground water. The advantages of ground water are threefold: the supply is consistent, it is typically free of environmental contaminants, and wild fish cannot enter the production facility from the water source.

Before having a well installed, it is a good idea to check with the U.S. Geological Survey office or local well permitting authority. This agency provides information on the principle aquifers in the area, chemical characteristics of the water, and the potential for water quality or quantity changes over time. These potential changes include saltwater encroachment and contamination resulting from surface land use practices.

An initial water quality analysis will ensure that the well water meets the biological requirements of the fish you plan to culture and that there are no contaminants that might affect the quality of the final product. Wells should be periodically checked to monitor water quality changes.

Surface Water

Pumping surface water from a ditch to a fish pond

Surface water is often used for fish production in areas where the terrain makes this possible, or where ground water is unavailable. Surface water sources include rainfall runoff that can be supplemented by ground water or springs, and streams, lakes and reservoirs.

In all cases, the watershed through which rainfall flows into the receiving source (stream, lake or pond) will greatly influence the quality of water available. The possibility of contamination by pollutants or undesirable wild fish stocks is increased by the use of surface water. Protecting the watershed can prevent contaminants in watershed production ponds. Potential contamination can be prevented by:

1. **Avoiding livestock feedlots**
2. **Avoiding industrial or active mining sites**
3. **Using only approved pesticides, according to label instructions, in orchards, row crops or pastures**
4. **Avoiding residential runoff, including that from septic tanks**
5. **Avoiding runoff from roadways or railroads.**

Surface waters are also subject to variation due to weather events. In dry years, the amount of water available may be greatly reduced. During and after rainfall events the waters can become muddied, thus increasing the amount of silt brought into the production ponds.

In many cases, surface water is preferred over ground water for economic reasons. In some areas, drilling a well to the depth required to obtain an adequate volume of the desired quality water is too costly, and alternative water sources from the surface are required. In all cases, if suitable quantities of ground water are available economically, this is the PREFERRED and recommended water source.

If you are forced to use surface waters, be aware of the possible dangers and take every precaution to ensure that your production facility is not compromised by unwanted contamination that can occur due to circumstances beyond your control.

Soil Type

The most important soil characteristic is water-holding capacity. The desired soils for pond construction usually contain more than 50 percent clay particles. These can be sandy clay, silty clay, clay loam, or heavy clay soils. If you are unsure what kind of soils are on the land you are considering, several resources can help you determine your soil types and suitability. The USDA Natural Resources Conservation Service (NRCS) can assist you with many aspects of planning and designing ponds. They can provide assistance in soil assessment as well as in layout design.

If you determine that your proposed site does not possess acceptable soils, you should seriously consider ANOTHER site. Although it is possible to make soil amendments to increase water holding capacity, it is usually also expensive and often inefficient to do so. In some cases, you can spend large sums of money on soil amendments and still have leaky ponds!

Topography and Proximity to Other Resources

The slope and lay of the land at and surrounding your production site are important influences on the type and design of facility you can establish. When a pond system is proposed, flat or gently sloping land is preferred, since the amount of dirt to be moved and the cost of construction is minimized.

Raceways are ideally suited for steep topography
Photo by Gary Fornshell

Extremely low-lying areas, however, should be avoided if they are subject to flooding either from rainfall runoff in the watershed or from encroaching waters from nearby streams or rivers. High waters not only bring in wild fish, which can be economically devastating, but they also contribute to erosion of levees, which can be a serious problem. Additionally, during periods of heavy rainfall, YOUR fish can escape from the pond, leaving you with only a pond of water and unmarketable wild fish to show for your hard work!

Utility access is important in site selection

It is also important that ponds be located so they can be drained by gravity flow. This is only possible if the surrounding land (or ditch) is lower than the bottom of the ponds. If the natural slope of the land does not allow gravity drainage, additional construction costs will be increased because ditches must be dug and maintained.

Access to the production facility is a key consideration. The ideal location is near primary roads or highways that allows transportation of products and supplies to and from the farm without unusual expense. All-weather, year-round access to the ponds or production complex is important. Feeding, harvesting, stocking, receiving feed and other supplies, general management and observation all require access to the facility, regardless of the weather. Thus, all-weather surfaces on the levees or roads are vital, and the proximity to public roads and travel routes is also important. It is of little benefit to produce a crop that cannot be transported to market due to inaccessible road conditions!

Another important consideration is proximity to power lines, telephone service and other utilities. In almost every conceivable production system, electrical power is an absolute necessity. Aerators powered by electricity are standard equipment on almost every kind of pond, and power to run recirculating systems, pumps and aerators will be required regardless of the production system that you use. It can be costly to run electrical service to remote locations, so before deciding on a particular site, check with the local power service provider to determine availability of electrical service, costs to install at the desired location, and standard or variable rates to which your farm will be subjected.

Telephone service considerations have become less critical with the advent of cellular communications, but even cellular coverage is not universal, and thus it is imperative to assess the availability of telephone service. Telephones are used not only for routine or emergency conversations, business transactions and general communication, but also increasingly for data transmission, either via the Internet, or through use of local monitoring systems that communicate to personal computers or to your telephone when transmitting routine or emergency data. Some automated monitoring systems use the telephone system to communicate with a PC to record data, as well as communicating with farm management to warn of problems or impending problems.

After determining that the chosen site is suitable and free from potential contaminants, the environmental impacts of the proposed facility are examined. Construction of ponds or other facilities in certain federally classified wetlands requires adherence to a permitting process administered through the U.S. Army Corps of Engineers and conformity to provisions of the Wetlands Act. Wetlands are defined as, "those areas that are inundated or saturated by surface or groundwater at a frequency and duration sufficient to support, and that under normal circumstances do support, a prevalence of vegetation typically adapted for life in saturated soil conditions. Wetlands generally include swamps, marshes, bogs and similar areas." In many cases wetlands considerations will prevent construction of facilities.

The Natural Resources Conservation Service provides technical guidance in assessing sites and determination of wetlands or other regulatory concerns. In some states a permit is required to drill wells, and there may be regulations concerning effluent discharge from the facility. Potential producers must determine the necessary regulatory requirements and coordinate with the appropriate agencies.

Pesticides

Avoid pesticide application near ponds

If the prospective site is located in an agricultural or former agricultural area, or in a location where pesticides are known to have either been stored or applied, it is highly advisable to check the soils for contamination. Highly suspect areas include any crop fields, old barn or storage sites, airplane runways and loading areas, field equipment wash-down areas, and any other location that might have accumulated pesticides over the years.

Pesticides in the group known as chlorinated hydrocarbons are especially persistent in the environment, and were used extensively in the 1940-1970 period, especially in cotton producing areas. If possible, check the land use history of the proposed site to determine which crops were grown and what other uses the land may have had in the past. This knowledge will provide useful insight as to the possible contaminants; thus reducing analytical costs, since there will be a specifically targeted group of suspected contaminants for which to examine.

If you find that your surface soil contains undesirable levels of pesticides, it is wise to take additional soil samples at a lower soil depth. If these deeper soils are clean, the contaminated surface soils can be removed and used as fill dirt on OUTSIDES of levees, where there is no contact with production water.

Since fish are typically grown in agricultural regions, great care should be taken to locate ponds in areas that eliminate or minimize the possibility of pesticide and chemical drift from other agricultural operations. Be careful to identify adjacent landowners and farm operators, and to communicate closely with these neighbors about the mutual need for coordination of activities, especially pesticide application. Agricultural operators adjacent to fish production facilities should always confer with neighboring fish farmers about scheduled pesticide applications, and mutually agree upon application methods, timing, etc.

Geographic Location

The location of the facility relative to other agricultural, industrial, civic or private land uses will greatly influence the suitability of the proposed site. Any activity that may impact the air, soil, or water resources on your farm will also potentially affect the operation of your production efforts. For example, if you are growing fish in a watershed pond and a land-clearing operation exists upstream, your facility is subject to significant inflow of mud and debris unless preventive measures are taken. Likewise, any pollutant upstream, or seeping into ground water supplies, will also become a factor in your production success or failure.

Geographic location also influences the types of products that can be feasibly cultured. Some species are adapted to a very narrow range of conditions, while others are tolerant and thrive in a wider range of environmental circumstances. Water quality and quantity, soil chemistry and temperature all will vary with location, and the range of variation can be quite high site to site. Geography has its greatest impact upon water temperature, and U.S. aquaculture consists of two general temperature-related systems: cool/cold water, and warm-water aquaculture.

Stocking

8

Regardless of the species you culture, you must stock the production units appropriately to produce the desired product. If you are producing fingerlings, larvae or other juvenile stages, stocking may consist primarily of broodfish that produce eggs and young, which may then be stocked into rearing units to grow to market size.

Most aquaculturists grow juvenile fish into market-sized adults. This involves stocking small fish, usually called fingerlings, into rearing units where they are fed and managed until they reach the appropriate size for market

Stocking a pond with bass fingerlings

The densities at which fish are stocked vary greatly by species and by type of production system. Stocking densities depend on several factors, all of which combine to compose what is called "carrying capacity." Carrying capacity is the total biomass or weight of fish that a given production unit can support. In most cases, aquaculture units are stocked at rates that greatly exceed "natural" carrying capacity. This "artificial" carrying capacity is established by the addition of supplemental feeds, aeration and other management inputs.

To determine the number of fish needed for stocking a production unit, multiply the number of fish desired per unit volume or area times the unit area or volume of the rearing facility. For example, if you want to stock 5,000 catfish per surface acre in a 15 acre pond, the total number of fish needed is:

Number to stock = stocking rate/acre X surface acres

= 5,000/acre X 15 acres

Number to stock = 75,000 fish

Some species are counted indirectly by using a sample count method. Sample counts can be done several ways. For example, suppose you want to stock 15,000 juvenile prawns per acre in a 3 acre pond. Rather than count out 45,000 prawns individually, an easier method would be to weigh out a specific poundage of the prawns, and then count the number of prawns per pound. Suppose you weigh out 1 pound of juvenile prawns, and then determine that there are 500 prawns in that 1 pound sample. Using this information, you can easily determine the weight per 1,000 prawns:

lb/1,000 prawns = 1,000/(# prawns/lb)

= 1,000/500

lb/1,000 prawns = 2

Then, to determine the number of pounds of prawns needed to stock the 3 acre pond:

lbs needed = number of prawns needed (2 lb/1,000 prawns)

= 45,000 X 2/1,000

= 90 lbs needed.

The actual stocking density used will depend upon the species and its environmental requirements. Water characteristics such as temperature, dissolved oxygen and the many other parameters that describe the water will play a major role in determining stocking density. To a great extent, stocking rates will be dependent on the degree of management inputs that are available from the manager.

The time of stocking production units can vary by species, type of unit or system, and the timing desired for production of the market-sized animals. For some species and production system types, stocking may be based on seasonal or other cyclic occurrences or timing. For others, time or year makes no difference, and the timing of stocking may depend more on market strategy or the availability of stocker size juveniles.

Feeding

9

Feeding is the single most critical, and often most expensive, component of managing an aquaculture operation. In almost all cases except the most extensive types of operations, feeding with a supplemental, manufactured feed is necessary to assure rapid, consistent growth of the crop. These feeds are usually nutritionally complete rations that have been developed through research and manufactured under strict control standards to ensure quality and consistency.

Automatic feeder over raceways used for tilapia production

In many operations, feed costs can represent 50 percent or more of the total annual operating costs. For this reason, it is important to use the proper feed and to take measures to be sure that the conversion of feed to fish flesh is efficient. This value, called Feed Conversion Ratio (FCR), is expressed as the number of pounds of feed required to produce one pound of flesh. Depending on the species being grown, it may take 1.5 to 5 pounds of feed to produce 1 pound of fish feed.

In selecting a proper feed, it is important to match the feed size to the size of the fish being fed. Smaller particle or pellet size is required for smaller fish, while larger pellets will be more efficiently used by larger fish.

The feed must be suitable to meet the nutritional needs of the species being cultured. A commercially manufactured feed is available for most species cultured in the U.S., and you can use these commercially available feeds with confidence.

Feeding rates will vary also by species and production system, but a few common factors contribute to determining these rates:

- **Water temperature**
- **Water quality**
- **Size of the feed particles**
- **Palatability of the feed to the fish**

The size of feed pellets should match the size of fish being fed
Photo by Jimmy Avery

- **Frequency of feeding**
- **Technique for feed delivery**
- **Location of feeding sites**
- **Type of feed used (i.e., floating, sinking, etc.)**

Feed can be delivered in a variety of ways, including feeding by hand, using automated, stationary feeders, using mobile feeders, or by allowing the fish to feed themselves using "demand" feeders. The choice of feed delivery system used will be dictated by logistical needs, resources available, amount of feed to be delivered, number of units to be fed daily, and the size, scope and type of operation.

Hand feeding fish in a cage system
Photo by Jimmy Avery

Water Quality

10

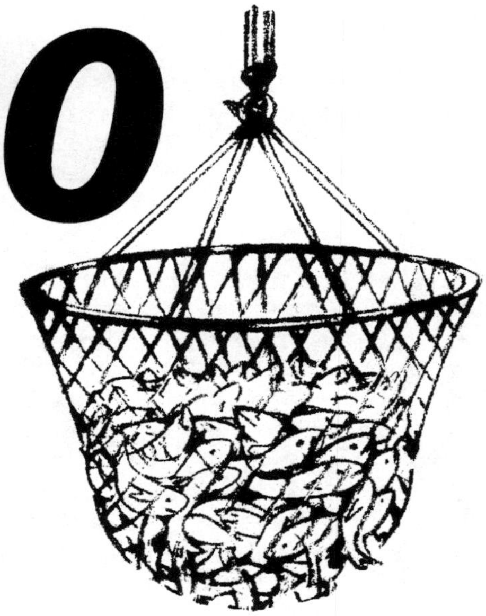

Because fish and other aquatic organisms are totally dependent upon the water surrounding them and are unable to escape this environment, the quality of the water is of great importance. Several water quality parameters are important in fish culture, including dissolved oxygen (DO), temperature, mineral content, and the concentrations of several waste products excreted by the organisms. The presence of fish or other aquatic organisms can also influence water chemistry. For example, fish excrete waste products that are high in ammonia and other nitrogen compounds, which tend to pollute the water. Likewise, plants can favorably influence water chemistry by adding oxygen and by removing nitrogen compounds. Consequently, the relationship between the cultured species and the surrounding water is a dynamic and crucially important one.

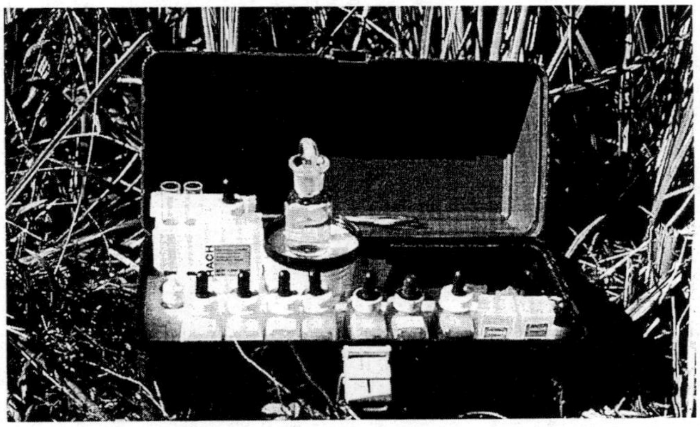

Water quality test kit

Dissolved oxygen (DO) is commonly considered the water quality factor of greatest importance to aquaculturists. Pure oxygen, dissolved in water, is absolutely necessary for aquatic organisms. All animals and plants require this basic element for survival, and die in a very short time when deprived of oxygen.

Oxygen management is critical not only to ensure adequate oxygen for survival, but also to ensure adequate oxygen to prevent the stress that organisms experience at levels that are lower than optimum. When fish are exposed for long periods of time to these low, non-lethal levels of DO, they may not die immediately, but they are much more likely to succumb to disease or other factors resulting in poor performance or ultimately, death.

Chronic exposure of aquatic animals to sub-lethal levels of DO results in reduced feeding, increased disease incidence, reduced feed conversion and subsequent poor production and reduced profits.

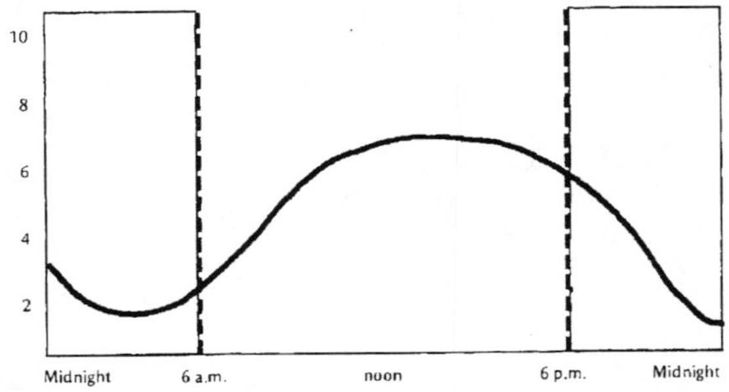

Twenty-four hour oxygen cycle in ponds

The minimum DO level that fish can tolerate varies by species and by temperature, but as a general rule, a minimum DO level of 4 parts per million (ppm) is desired. Cool and cold water species are less tolerant of low DO than most warm water species.

The primary source of oxygen in freshwater is the process of photosynthesis, whereby plants use energy from the sun, carbon dioxide (CO_2), and water to produce starches, which they use as food. In the process, an excess amount of oxygen is produced, resulting in increased DO levels in the water following photosynthesis. Because of photosynthesis, DO levels can vary greatly over a 24-hour period. All organisms in the pond use oxygen continuously (respiration), but a surplus of oxygen is produced only during daylight hours. Thus, oxygen levels tend to be highest in the late afternoon, and tend to be lowest just before dawn. The surplus DO that is produced during the day must be adequate to meet the needs of the pond and its inhabitants throughout the following night until daylight the next morning. If the surplus is not adequate, oxygen depletion will result.

Solubility of oxygen in parts per million (ppm) in fresh water at various temperatures and at a pressure of 760 mm Hg (sea level).

Temperature °F	Temperature °C	Concentration of oxygen in ppm	Temperature °F	Temperature °C	Concentration of oxygen in ppm
32	0	14.6	69.8	21	9.0
33.8	1	14.2	71.6	22	8.8
35.6	2	13.8	73.4	23	8.7
37.4	3	13.5	75.2	24	8.5
39.2	4	13.1	77	25	8.4
41	5	12.8	78.8	26	8.2
42.8	6	12.5	80.6	27	8.1
44.6	7	12.2	82.4	28	7.9
46.4	8	11.9	84.2	29	7.8
48.2	9	11.6	86	30	7.6
50	10	11.3	87.8	31	7.5
51.8	11	11.1	89.6	32	7.4
53.6	12	10.8	91.4	33	7.3
55.4	13	10.6	93.2	34	7.2
57.2	14	10.4	95.0	35	7.1
59	15	10.2	96.8	36	7.0
60.8	16	10.0	98.6	37	6.8
62.6	17	9.7	100.4	38	6.7
64.4	18	9.5	102.2	39	6.6
66.2	19	9.4	104.0	40	6.5
68	20	9.2			

Solubility of oxygen table

The amount of oxygen that water can hold depends upon three factors: atmospheric pressure, salinity, and temperature. As altitude increases, water can hold less oxygen. Salinity is not a factor in most freshwater production systems. Thus, the most important influence on the solubility of oxygen in water is TEMPERATURE. As temperature increases, the solubility of oxygen in water decreases; in effect, warm water can hold less oxygen than cool water. This relationship between temperature and DO causes oxygen problems in ponds, especially when water is warm or hot. Although a DO problem can occur at any temperature, warm temperatures typically pose a greater management challenge to producers.

An additional complicating factor in oxygen management in warm water is that plants and animals need and use more oxygen in warm water than in cool water. Thus, the greatest oxygen demand occurs at a time when DO concentrations tend to be lower. Additionally, feeding rates are usually higher in warm water, and warm water species grow faster at higher temperatures. Increased feeding rates, and increased digestive activity result in higher consumption of oxygen.

Oxygen depletions

Oxygen problems or depletions can be caused by several factors.

Any time consumption of oxygen exceeds production, depletion may occur. For example, as fish are fed heavily during warm weather, microscopic algae populations build up to high levels. As the density of these algae increase, they crowd close to the surface and shade the lower water reaches. Most of the oxygen produced is in the upper layers of water, with little oxygen in lower depths. Eventually, if left unchecked, the consumption rates for DO will exceed the amount produced in the upper levels and depletion can occur.

DO depletion can also occur if the algae suddenly die. Algae can die because they have depleted their food source, if they do not receive enough sunlight to conduct photosynthesis or due to application of herbicides in the system. Regardless of the reason for death, once the algae die, the net surplus of oxygen they HAD been producing is quickly used up by the decay process that follows death. With no other source of surplus oxygen, depletion will soon follow.

Algae die-offs are easy to diagnose. Usually, the water will turn from a "green" color (color of the algae) to a mottled or streaky gray, brown or black when the algae begin to die. Often a "sewage" type smell is evident. When this occurs, a depletion is imminent unless aeration is provided.

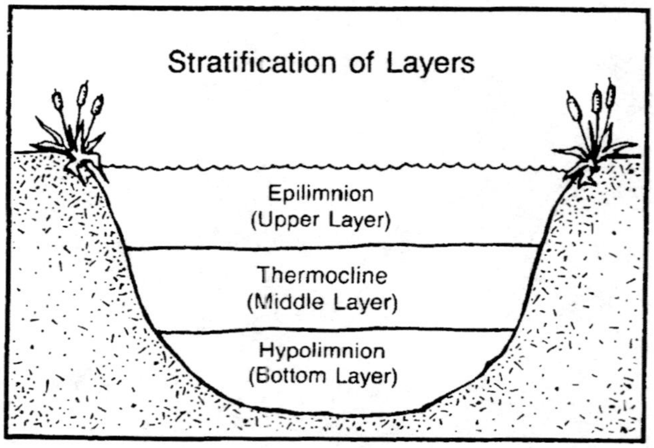

Stratification of layers in a pond

Another kind of oxygen depletion can occur called a "turnover". This occurs when the pond is allowed to "stratify" due to temperature differences between surface and bottom waters. Warm surface waters are lighter than cooler bottom waters, and a distinct layering, or stratification can establish in a pond if these temperature differences develop. The light, warm surface waters usually have adequate DO since this is where photosynthesis takes place, but the cooler bottom waters may have very little DO.

If this layering of water lasts for a significant amount of time, eventually the oxygen in the lower water will be used up. If the layers are mixed, the DO that is found in the upper layer is redistributed throughout the water column, resulting in lower DO concentrations and a possible DO depletion. The layers can be mixed by wind action, aeration, or by anything that may cool the top water down, making them heavy and causing them to sink. This can occur after a rainfall event, or when the air temperature suddenly cools after a warm period.

Monitoring DO

Dissolved oxygen can be easily and accurately measured using either an inexpensive chemical test kit, or a more expensive but more accurate and reliable electronic meter. Chemical kits are acceptable if the number of ponds, or units, is small, but if you must make multiple DO checks routinely, investment in an electronic meter is wise. Meters provide fast, reliable readings, and allow you to check DO when you need to without undue time investment.

During warm months, or in intensive culture systems, DO should be checked at least twice daily and the results recorded. Dawn and dusk are the two most critical times to check DO, but in some cases it will be necessary to check throughout the day and night at various intervals, especially if unstable conditions exist that may result in rapid loss of DO.

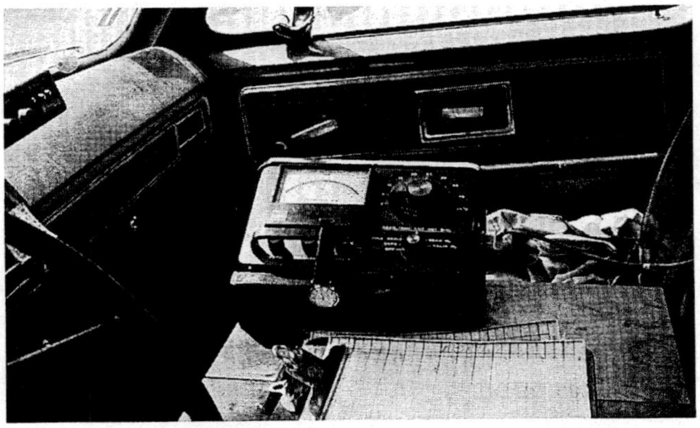

DO meter

If DO records are kept, you can often predict potential depletions and thus avoid problems by taking corrective action BEFORE a problem occurs. For example, periodic review of recent DO levels at dawn and dusk can provide insight as to trends in DO levels. If morning DO appears on a steady decline over several days, this may be an indication that corrective action may be appropriate to avoid dangerously low DO levels in the near future.

Another method that is commonly used is a simple graphical tool that works nicely in most cases. This method is based on the fact that declines in DO in pond systems during the night is usually a straight line over time. The technique used is to:

1. Measure the DO at dusk and plot this point on a graph.

2. Measure DO again 2-3 hours later and plot on the same graph.

3. Draw a straight line between these two points and extend it to a point where it crosses the 4 ppm DO concentration mark.

4. Look at the time that corresponds with this intersection and you will have a good idea as to whether aeration will be required before sunrise the next morning.

Using this method will provide opportunity to be proactive in dealing with potential problems BEFORE they occur. Bear in mind that this technique is not fool-proof, and does not necessarily apply in all situations, but it can be a useful tool in your management arsenal. No method or "trick" can be used to replace diligence and close monitoring of your system and the animals within that system.

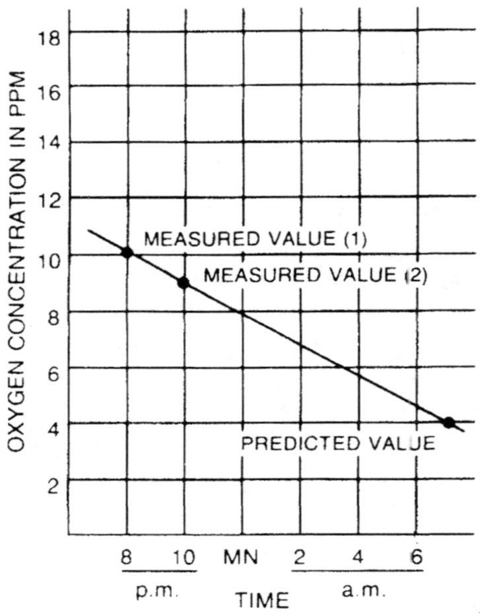

Graphic method of predicting nighttime oxygen depletions in catfish pond. In this example, it is predicted that no problem will develop in the pond during the night.

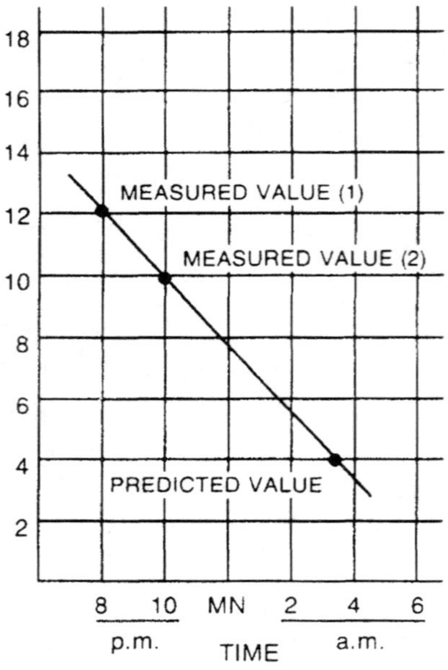

Graphic method of predicting nighttime oxygen depletions in catfish pond. In this example, it is predicted that the oxygen concentration will drop to 4 ppm by about 3:30 a.m., thus indicating possible corrective measures should be taken about 3:00 a.m.

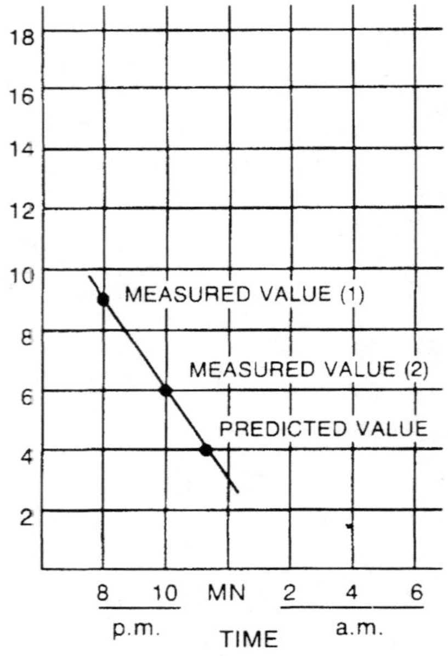

Graphic method of predicting nighttime oxygen depletions in catfish pond. In this example, it is predicted that the oxygen concentration will drop to 4 ppm by about 11:30 p.m. and unless emergency measures are taken a severe fish kill will probably occur at about 1:00 a.m.

Graphic method of predicting nighttime oxygen depletions in catfish pond

Correcting DO Problems

Dissolved oxygen problems are corrected by adding oxygen. Although pure oxygen can be added through diffusion from containers, this method is expensive and usually inefficient in all but the most intensive systems. A better alternative is to take advantage of the high concentration of oxygen in the atmosphere by diffusing it into the water.

Mechanical aeration has proven to be a cost effective and efficient means of adding oxygen to the water, and many different types of mechanical aeration devices are used. These include paddlewheels, surface sprayers, pump sprayers, diffused air systems, propeller-aspirators, and gravity aerators. The choice of aerator design(s) depends upon the type of production system, energy availability, other equipment availability, and other logistical factors that may be unique farm-to-farm.

Gravity aerator

PTO driven paddlewheel aerator in use

Paddlewheel aerators have been in use for many years. They can be either electric powered, powered by self-contained engines, or powered by a power-take-off (PTO) from a farm tractor.

In pond systems, most producers have come to rely on electric or diesel driven, fixed position paddlewheels for routine or scheduled aeration, and upon PTO aerators for emergency or supplemental aeration. Self-contained units are on floats and fixed by steel rods or cables to the bank in a stationary position. PTO units can be pulled by tractor from pond to pond, making them ideal for emergency use. A disadvantage to paddlewheels is that they tend to erode the pond bottom below and in front of the aerator.

Surface spray pumps operate with a submersible electric motor that rotates an impeller to pump surface water into the air as a spray. These are floating units, usually lightweight and portable. Spray pumps can be used continuously, and can be effective in small ponds or other units. They are not efficient in larger ponds since they do not move much water and the oxygen transfer rates are low.

Bankwasher powered by a diesel engine

Pump sprayer aerators are used on most farm operations using ponds to grow fish. They are powered either by PTO, electricity or diesel. These are equipped with either an impeller suction pump, and impeller lift pump or a turbine pump, and discharge water various ways. These are often called bankwashers, since they discharge water parallel to the bank, and spray water through constrictions in the pipe into the air. Bankwashers do not erode the pond bottom like paddlewheels, and are thus preferred by many producers.

Diffused air systems operate using low pressure air blowers or compressors that force air through aeration lines or diffuser stones, releasing air bubbles below the surface of the water. These types of systems are not typically used in larger ponds, but are widely used in smaller production units such as tanks and raceways.

Propeller-aspirator pumps use a rotating, hollow shaft attached to a motor shaft. The submerged end of the rotating shaft is fitted with an impeller that pulls air down the hollow shaft. Air passes through a diffuser and enters the water as fine bubbles that are mixed into the pond water by the turbulence created by the propeller. These are powered by electricity, and usually used in small ponds or tanks.

Placement of Aerators

Location of aerators is most critical in ponds, especially larger ones. It is important to locate the aerator where maximum transfer of oxygen into a maximum pond area can occur. It is also important to have enough aeration so that fish can get fresh, oxygenated water from anywhere in the pond or system without undue stress. For fixed position paddlewheel or bankwasher aerators in ponds, the ideal location is mid-way along the longest side, with the discharge directed toward the middle of the pond. This provides the most efficient water circulation patterns, but also may result in higher costs to run power to mid-points on long levees.

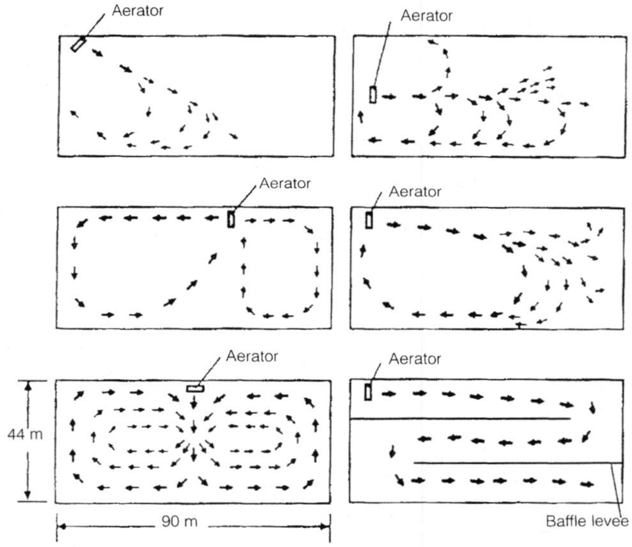

Water circulation patterns for various placements
Adapted from Boyd, 1989

It is important to use enough aeration power to adequately serve the water volume. This will often mean using multiple aerators per pond, especially if the aerators are small. If multiple aerators are used, locate the aerators in relatively close proximity to each other. This will allow fish to move quickly and without stress from one aerated area to another if one unit suddenly fails. During a DO depletion, you DO NOT want fish trying to swim long distances through poorly oxygenated water to get to an aerated area. The stress will compromise the fish and make a bad situation worse. With some animals, movement toward an aerated zone is difficult or impossible, so it is critical to aerate as much of the pond area as possible.

Nitrite

Nitrite enters a fish culture system after fish digest feed and the excess nitrogen is converted into ammonia, which is then excreted as waste into the water. Total ammonia nitrogen (TAN; NH_3 and NH_4+) is then converted to nitrite (NO_2) which, under normal conditions, is quickly converted to non-toxic nitrate (NO_3) by naturally occurring bacteria. Uneaten (wasted) feed and other organic material also break down into ammonia, nitrite and nitrate in a similar manner.

Brown blood disease occurs in fish when water contains high nitrite concentrations. Nitrite enters the bloodstream through the gills and turns the blood to a chocolate-brown color. Hemoglobin, which transports oxygen in the blood, combines with nitrite to form methemoglobin, which is incapable of oxygen transport. Brown blood cannot carry sufficient amounts of oxygen, and affected fish can suffocate despite adequate oxygen concentration in the water. This accounts for the gasping behavior often observed in fish with brown blood disease, even when oxygen levels are relatively high.

Nitrite problems are typically more likely in closed, intensive culture systems due to insufficient, inefficient or malfunctioning filtration systems. High nitrite concentrations in ponds occur more frequently in the fall and spring when temperatures are fluctuating, resulting in the breakdown of the nitrogen cycle due to decreased plankton and/or bacterial activity.

A reduction in plankton activity in ponds (due to lower temperatures, nutrient depletion, cloudy weather, herbicide treatments, etc.) can result in less ammonia assimilated by the algae, thus increasing the load on the nitrifying bacteria. If nitrite levels exceed that which resident bacteria can rapidly convert to nitrate, a buildup of nitrite occurs, and brown blood disease is a risk. Although nitrite is seldom a problem in systems with high water exchange rates or good filtration, systems should be monitored year-round and managed when necessary, to prevent severe economic loss from brown blood in any fish culture facility.

Susceptibility of Fish Species to Brown Blood Disease

Largemouth and smallmouth bass, as well as bluegill and green sunfish, are resistant to high nitrite concentrations.

Catfish and tilapia, are fairly sensitive to nitrite, while trout and other coolwater fish are sensitive to extremely small amounts of nitrite. Goldfish and fathead minnows fall in between catfish and bass in their susceptibility to brown blood disease resulting from high nitrite levels. Striped bass and its hybrids appear sensitive to nitrite, but little is known about the relative sensitivity compared to other species.

Treatment and Prevention

Since this is a nitrogen-related problem, the most obvious preventive measure is to reduce or minimize the amount of nitrogen incorporated into the system by reducing feeding rates. However, in a modern intensive pond or closed system fish culture with high densities and rapid growout, most farmers do not consider long-term feed reduction a viable option. Luckily, although we often cannot prevent the occurrence of high nitrite, its effects can be minimized or neutralized safely and economically.

Sodium chloride (common salt; NaCl) is used to "treat" brown blood disease. Calcium chloride can also be used but is typically more expensive. The chloride portion of salt competes with nitrite for absorption through the gills. Maintaining at least a 10 to 1 ratio of chloride to nitrite in a pond effectively prevents nitrite from entering catfish.

Where catfish (or other fish) have bacterial and/or parasite diseases, their sensitivity to nitrite may be greater, and a higher chloride to nitrite ratio may be needed to afford added protection from nitrite invasion into the bloodstream. As a general rule, catfish producers strive to maintain at least 75-100 ppm chloride in pond waters as "insurance" against high spikes of nitrite concentration. It is likely a good idea for culturists of other species to assume that nitrite is a potential problem and use salt as an insurance buffer as well.

How to Calculate the Amount of Salt Needed

Before treatment rates can be calculated, chloride and nitrite concentrations in the water, as well as pond or tank volume, must be determined. Commercially available water quality test kits can be used. Contact your extension fisheries or aquaculture specialist for assistance in locating sources for test kits and conducting and interpreting these tests.

The amount of salt needed for the pond can be calculated using the following formulas:

FORMULA 1

(10 x pond nitrite concentration) - (pond chloride concentration) = parts per million (ppm) of chloride to add to the pond.

The number "10" used in this formula is the minimum desired chloride:nitrite ratio number. It is used here to get a 10 to 1 chloride to nitrite ratio. If a higher ratio is desired, substitute the 10 with the higher number.

If the answer is zero or a negative number, chloride concentration is sufficient to prevent brown blood disease.

Use the answer from Formula 1 above in the following formula:

FORMULA 2

Surface acres x average depth in feet x ppm of chloride to add to the pond x 4.5 = pounds of salt (NaCl) needed to add to the pond.

4.5 pounds of salt are needed to increase the chloride concentration by 1 ppm in an acre-foot of water.

Example

The following readings are obtained from a 20-acre catfish pond with an average depth of 4 feet:

4 ppm nitrite

15 ppm chloride

Use Formula 1:

(10 x 4 ppm nitrite) - 15 ppm chloride in the pond = 40 - 15

= 25 ppm chloride to add to the pond

Now use Formula 2:

20 acres x 4 feet average depth x 25 ppm chloride to add to the pond x 4.5 = 9,000 pounds of salt needed to add to the pond.

Application of Salt

Distribute the salt evenly and quickly when fish are suffering from brown blood disease. Farmers have used feed trucks, airplanes, paddle wheels and front-end loaders to distribute salt. It takes about 24 hours after salt is applied to a pond for the brown blood condition to be alleviated.

A good water quality monitoring program can help prevent brown blood disease. Pond water should be checked for nitrite 2 to 3 times a week during fall and spring, and at least weekly the remainder of the year. A chloride to nitrite ratio of at least 10:1 is recommended.

Check ponds daily during a known high nitrite incident, even if adequate chlorides are in the ponds. Also check chloride after periods of heavy rain or active flushing from well water; both these events can dilute chloride concentrations and reduce the chloride:nitrite ratio.

Nitrite can increase very suddenly, so it is advisable to keep a 75 to 100 ppm chloride concentration at all times to act as a buffer when nitrite suddenly increases. This is a standard practice in the catfish industry, and incidents of brown blood disease in catfish ponds have become very rare. As an example, if your water has a chloride concentration of 4 ppm and you want to increase it to 75 ppm, simply add 71 ppm chloride to your pond: use Formula 2 to calculate the pounds of salt needed.

Another way to help prevent brown blood disease is by checking total ammonia nitrogen (TAN) concentrations in ponds every week. Every 1 ppm TAN can convert to 3 ppm nitrite in a relatively short period. High TAN levels can alert the farmer to anticipate nitrite problems within a few days, and nitrite problems can thus be predicted and then prevented.

In many areas, water contains high natural concentrations of chloride, and addition of salt as "insurance" is not needed. Water should still be monitored frequently, however, since chloride levels can fluctuate widely and frequently.

Outlook

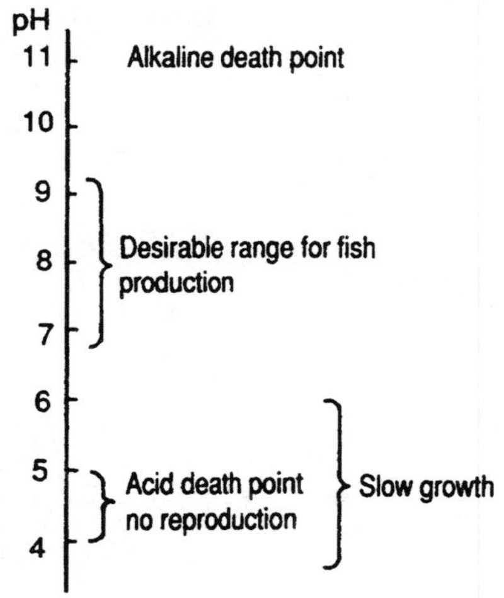

pH ranges for fish production

Brown blood disease can be prevented, or at least minimized, by close monitoring of nitrite, chloride and TAN, and by maintaining the proper chloride to nitrite ratio. If brown blood disease does occur, adding salt to the water can reverse the condition. Catfish (and likely other fish) surviving brown blood disease or nitrite stress are more susceptible to bacterial infections, anemia (white-lip or no-blood), and other stress-related diseases. These secondary problems, such as Aeromonas, Columnaris, etc., infections often occur 1 to 3 weeks after brown blood disease occurs.

Research is currently underway to determine whether even higher levels of chloride may be beneficial in reducing sub-lethal, chronic stress on fish from nitrite or other stressing factors.

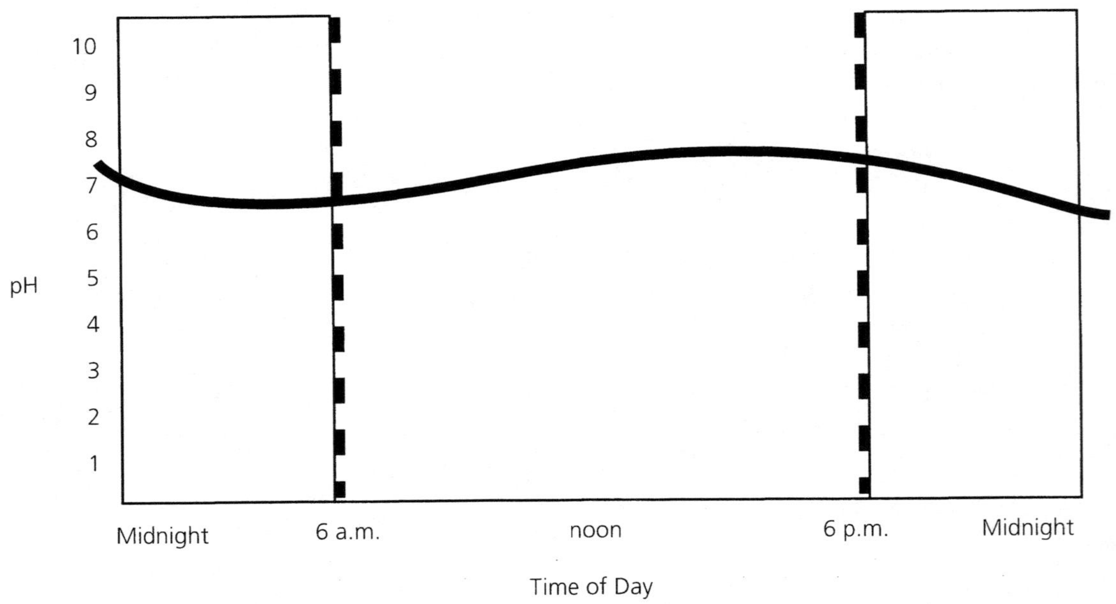

Twenty-four hour pH cycle in ponds

pH

The water quality parameter known as pH is a numerical value that ranges from 0-14, and indicates how acid or basic water is at a given time. Technically, pH is defined as the negative logarithm of the hydrogen ion concentration, but for practical purposes it can simply be considered an index to how acid or non-acid water is. A pH of 7 indicates that water is "neutral, that is neither acid nor basic. Values below 7.0 indicate that water is acid, and has more hydrogen (H^+) ions than hydroxyl (OH^-) ions, whereas a pH above 7.0 means that water is basic and has more (OH^-) than (H^+) ions. The more acidic water becomes, the lower will be its pH.

The pH scale ranges from 0-14, with each unit change in pH representing a 10-fold increase in the H^+ ion concentration. The value of pH in a body of water at any given time is dependent on several factors and interactions among those factors. Some of these factors include the types of plant life present, the amount of photosynthesis being conducted, the basic chemical and mineral concentration of the water, and respiratory activity of the animals and plants living in the water.

Most cultured fish can tolerate pH between about 4 and 11, although many species are extremely sensitive to pH levels and to changes in pH. The optimum range for most aquaculture species is 6.5-9.0.

The pH of a body of water cycles on a 24-hour basis, in a fashion similar to the daily oxygen cycle. As aquatic plants, especially planktonic algae, conduct photosynthesis, they remove carbon dioxide (CO_2) from the water. Since CO_2 is an acid, removal of this compound results in a less acid water, with an increase in pH. However, as dusk approaches and photosynthesis slows and stops at sunset, CO_2 is no longer being removed from the system. Consequently, during the night, as organisms produce CO_2 through respiration, a surplus of CO_2 builds up until dawn, resulting in a decrease in pH.

Although pH is an important chemical factor affecting metabolic process, its daily fluctuating cycle remains rather consistent day-to-day, and thus is not usually closely monitored unless there is another problem that is affected by pH. The most common reason for close monitoring of pH is an episode of increased ammonia levels. In fact, any time an ammonia test is run, the pH must also be checked in order to calculate the percentage of toxic form of ammonia present.

The pH also affects the toxicity and efficiency of many compounds that may be used in fish culture, including some herbicides and drugs. Therefore, pH should be monitored routinely and you must be aware of the pH in your pond or tank before adding any water treatments, treating for disease or parasites, or any other activity of potential impact to your fish.

Total Alkalinity

Total alkalinity is the total concentration of bases in water expressed as parts per million (ppm) or milligrams per liter (mg/L) of calcium carbonate ($CaCO_3$). These bases are usually bicarbonates (HCO_3) and carbonates (CO_3), and they act as a buffer system that prevents drastic changes in pH. For example, in waters with low alkalinity, pH might fluctuate from 6 or lower to as high as 10 or above; while in high alkalinity waters, pH might fluctuate from about 7.5 to 8.5.

Greater production in bass-bluegill ponds is attained in high alkalinity waters because this pH buffering capacity makes phosphorus and other essential nutrients more available to the algae bloom.

Total alkalinity is not the same as hardness. Calcium (Ca^{++}) and magnesium (Mg^{++}) are primarily responsible for hardness. However, in most waters, alkalinity and hardness have similar values because the carbonates and bicarbonates responsible for total alkalinity are usually brought into the water in the form of calcium carbonate or magnesium carbonate. Waters with high total alkalinity are not always hard, since the carbonates can be brought into the water in the form of sodium or potassium carbonate.

Improving Alkalinity

The most common method of increasing alkalinity in waters is by adding agricultural limestone (calcium carbonate). Fish managers in Europe and Asia have reported that applications of lime to fish ponds on soils of low calcium content resulted in greater fish production. Fish managers in the United States have found that applications of lime to lakes increased pH and total alkalinity and also cleared water of brown stain from humic substances. Clear water results in deeper light penetration and greater photosynthesis, as well as increased zooplankton production.

Fertilization does not produce adequate phytoplankton (algae bloom) in many ponds with soft waters and acid muds because carbon dioxide is in short supply, and added phosphate adheres tightly to the bottom muds. Addition of lime to these ponds elevates total bicarbonate alkalinity and increases pH in the water. Neutralization of bottom muds with lime prevents phosphate from adhering to it, thereby increasing phosphorus concentrations in the water. As a result of these changes in water quality, phytoplankton blooms develop upon application of inorganic fertilizers. Liming soft water ponds to increase alkalinities to at least 20 ppm is beneficial to fish production.

As a general rule, ponds with alkalinities less than 20 ppm do not respond well to fertilization and should be limed in the fall before springtime fertilization is initiated. Liming ponds with alkalinities greater than 20 ppm typically will not increase the response to fertilizer. It does, however, increase the alkalinity and pH buffer capacity.

Copper Sulfate Treatment Dependent on Total Alkalinity

Copper sulfate ($CuSO_4$) is used commonly in fish culture ponds to treat for weeds and algae; it must be applied at a rate dependent on the total alkalinity of the water. At lower alkalinities, copper becomes more toxic and should not even be used at alkalinities below 50 ppm. Conversely, copper sulfate becomes ineffective at alkalinities above 300 ppm.

Fish can be killed if treatment rates in ppm of copper sulfate exceed total alkalinity divided by 100. On the other hand, treatments at lesser concentrations are ineffective. For effectiveness against weeds and algae, you must apply sufficient amounts of copper sulfate relative to the amount of carbonates in the water. If too much carbonate is present in the water, relative to the amount of copper sulfate added, the copper (Cu^{++}) will settle out as copper carbonate and will not be available in the water to treat the problem.

Total alkalinity divided by 100 is the rate in ppm at which copper sulfate should be applied.

Example: Water from a 15-acre pond (averaging 4 feet deep) measures 171 ppm total alkalinity. 171 ppm total alkalinity divided by 100 = 1.71 ppm copper sulfate to be used in this pond. This answer is then used in the following formula to determine pounds of copper sulfate to use:

Pond water acres x depth in feet x ppm copper sulfate x 2.72 pounds of copper sulfate per 1 acre-foot per 1 ppm.

The pounds of copper sulfate to use in this pond would be:

15 acres x 4 feet x 1.71 ppm copper sulfate x 2.72 = 279 pounds of copper sulfate.

Total Hardness

Total hardness is a measure of the sum total of titrable bases in the water. It indicates the concentrations of divalent metal ions (usually calcium and magnesium) in water. Total hardness is expressed in ppm of equivalent calcium carbonate.

The desirable range for total hardness is 20 ppm to 300 ppm. Levels below 20 ppm may cause problems in hatcheries, where developing eggs and fry need adequate calcium levels. Juveniles and adults of some species are sensitive to low hardness, and may require values of 100ppm or higher. Total hardness in excess of 300 ppm can also present management problems due to the effects of "hard" water on the species being cultured.

Low hardness levels are easily corrected using either agricultural lime or calcium chloride as a water treatment. Other materials can also increase total hardness levels, but may pose other risks or threats to the safety of the animals being cultured.

In many cases, total hardness values and total alkalinity values will be similar, but in many areas of the country this will not be true. Always be sure to test hardness and alkalinity independently so that you know how the two are related in your particular waters.

Ammonia

Ammonia is the major end product in the breakdown of proteins in fish. Fish digest the protein in their feed and excrete ammonia through their gills and in their feces. The amount of ammonia excreted by fish varies with the amount of feed put into the pond or culture system—increasing as feeding rates increase. Ammonia also enters the pond from bacterial decomposition of organic matter such as uneaten feed or dead algae and aquatic plants.

Forms and Toxicity

Total ammonia nitrogen (TAN) is composed of toxic (un-ionized) ammonia (NH3) and nontoxic (ionized) ammonia (NH+4). Only a fraction of the TAN exists as toxic (un-ionized) ammonia, and a balance exists between it and the nontoxic ionized ammonia:

H+ + (NH3) (NH+4)

The proportion of TAN in the toxic form increases as the temperature and pH of the water increase. For every pH increase of one unit, the amount of toxic un-ionized ammonia increases about 10 times. The amount of toxic un-ionized ammonia in your pond can be found by measuring the TAN with a water quality test kit and then looking up the fraction of TAN that is in the toxic form on Table 1, which is based on water temperature and pH. Multiply this fraction by the TAN to find the concentration (mg/L or ppm) of toxic un-ionized ammonia present in the water. For example, if water pH is 8.6, water temperature is 30oC, and TAN is 3 mg/1 (ppm), multiply 0.2422 (from Table 1) by 3 mg/1 (ppm) to obtain 0.73 mg/1 (ppm) toxic un-ionized ammonia.

Uptake (assimilation) of ammonia by plankton algae is important in reducing the amount of ammonia coming in contact with fish. Ammonia increases in the fall and winter due to reduced algae populations in the pond and algae populations which are not as capable of taking ammonia from the water. Additionally, lower water temperatures slow down aerobic bacterial activity, thus slowing the nitrification process whereby ammonia is converted to harmless nitrate. Algae die-offs can also lead to very high ammonia concentrations, but, fortunately, the low pH associated with the disappearance of the algae reduces the proportion of toxic un-ionized ammonia present.

Dangerous short-term levels of toxic un-ionized ammonia which are capable of killing fish over a few days start at about 0.6 mg/1 (ppm). Chronic exposure to toxic un-ionized ammonia levels as low as 0.06 mg/1 (ppm) can cause gill and kidney damage, reduction in growth, possible brain malfunctioning, and reduction in the oxygen-carrying capacity of the fish.

Treatments

Treatment for high TAN concentrations is difficult in large pond culture systems. Pumping fresh water into the pond is not a practical or economical means of reducing the ammonia level for the whole pond; it does, however, provide a small area near the inflowing water where fish can go to find some relief. Maintaining high dissolved oxygen levels by aeration will slightly reduce the toxic effect of un-ionized ammonia. Over time with higher DO, TAN levels may be reduced through increased aerobic bacterial activity. Temporary reduction of feeding rates is recommended until TAN levels decrease to an acceptable level.

Prevention of high TAN is a better approach to the problem. The use of lower feeding rates and good feeding practices plays a big role in keeping TAN levels low; problems with high TAN concentrations can be expected when feeding rates exceed 100 pounds per acre per day, or when excessive feed waste is occurring. Fish should not be overfed, and the feeder should be sure that fish are consuming feed offered. This is both practical and economically important, since feed costs are a major portion of production costs.

With pond and tank stocking densities continually increasing, it is not often considered economically practical to reduce feeding rates. However, the organic loading in these systems is a major factor that must be dealt with. Intensive recirculating systems may be better suited to handle these excessive amounts of nitrogen, but most pond systems likely have a finite limit to the amount of nitrogen and organic loading that can be managed. Unless more efficient management methods are developed, nitrogen and organic loading may become a limiting factor in stocking and production rates in culture ponds.

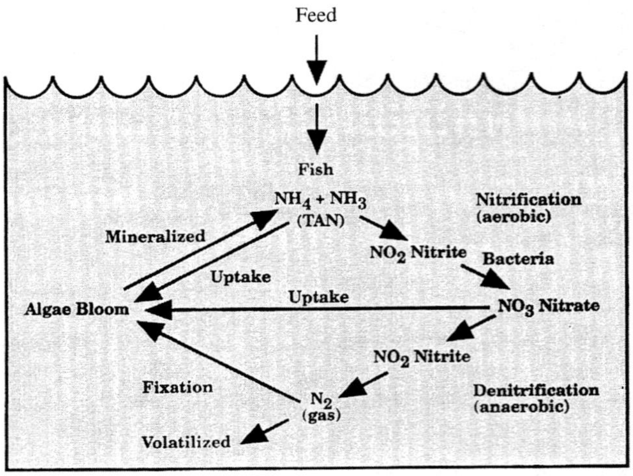

Nitrogen cycle in a fish pond

TABLE 1. Fraction of toxic (un-ionized) ammonia in aqueous solutions at different pH values and temperatures. Calculated from data in Emerson, et al (1975). To determine the amount of un-ionized ammonia present, get the fraction of ammonia that is in the un-ionized form for a specific pH and temperature from the table. Multiply this fraction by the total ammonia nitrogen present in a sample to get the concentration in ppm (mg/1) of toxic (un-ionized) ammonia.

Temperatures (oC)

pH	6	8	10	12	14	16	18	20	22	24	26	28	30
7.0	.0013	.0016	.0018	.0022	.0025	.0029	.0034	.0039	.0046	.0052	.0060	.0069	.0080
7.2	.0021	.0025	.0029	.0034	.0040	.0046	.0054	.0062	.0072	.0083	.0096	.0110	.0126
7.4	.0034	.0040	.0046	.0054	.0063	.0073	.0085	.0098	.0114	.0131	.0150	.0173	.0198
7.6	.0053	.0063	.0073	.0086	.0100	.0116	.0134	.0155	.0179	.0206	.0236	.0271	.0310
7.8	.0084	.0099	.0116	.0135	.0157	.0182	.0211	.0244	.0281	.0322	.0370	.0423	.0482
8.0	.0133	.0156	.0182	.0212	.0247	.0286	.0330	.0381	.0438	.0502	.0574	.0654	.0743
8.2	.0210	.0245	.0286	.0332	.0385	.0445	.0514	.0590	.0676	.0772	.0880	.0998	.1129
8.4	.0328	.0383	.0445	.0517	.0597	.0688	.0790	.0904	.1031	.1171	.1326	.1495	.1678
8.6	.0510	.0593	.0688	.0795	.0914	.1048	.1197	.1361	.1541	.1737	.1950	.2178	.2422
8.8	.0785	.0909	.1048	.1204	.1376	.1566	.1773	.1998	.2241	.2500	.2774	.3062	.3362
9.0	.1190	.1368	.1565	.1782	.2018	.2273	.2546	.2836	.3140	.3456	.3783	.4116	.4453
9.2	.1763	.2008	.2273	.2558	.2861	.3180	.3512	.3855	.4204	.4557	.4909	.5258	.5599
9.4	.2533	.2847	.3180	.3526	.3884	.4249	.4618	.4985	.5348	.5702	.6045	.6373	.6685
9.6	.3496	.3868	.4249	.4633	.5016	.5394	.5762	.6117	.6456	.6777	.7078	.7358	.7617
9.8	.4600	.5000	.5394	.5778	.6147	.6499	.6831	.7140	.7428	.7692	.7933	.8153	.8351
10.0	.5745	.6131	.6498	.6844	.7166	.7463	.7735	.7983	.8207	.8408	.8588	.8749	.8892
10.2	.6815	.7152	.7463	.7746	.8003	.8234	.8441	.8625	.8788	.8933	.9060	.9173	.9271

Source: Emerson, K., R. C. Russo, R. E. Lund, and R. V. Thurston. 1975. Aqueous ammonia equilibrium calculations: effect of pH and temperature.

Journal of the Fisheries Research Board of Canada. 32:2379-2383.

11 Fish Health and Disease Management

Stress Prevention and Its Role in Fish Health

There is no profit in a fish kill

Prevention, rather than treatment, should be the goal in fish health management. Providing a good environment may reduce disease incidence, assuring adequate nutrition, and preventing stresses related to overcrowding or improper handling. Unfortunately, economical and practical considerations often limit the extent to which stress can be minimized under commercial conditions. Fish producers must therefore use common sense and prevent situations that will lead to increased incidence or severity of disease.

Many of the disease organisms that affect fish are a natural part of the pond environment. Usually, they do not cause problems. There is a unique relationship among the fish, disease organisms, and the environment under normal environmental circumstances; that is, there is a delicate three-way balance among the fish, the disease organisms, and the environment. The fish's immune system typically is able to keep the disease in check. Any alteration, however, in one or more of the many environmental characteristics may result in a shift in the balance to the advantage or to the disadvantage of either the fish or the disease organism.

Environmental conditions directly affect the fish, often to the point of stress. Stress causes the fish to lose their ability to maintain the delicate balance amid environment and disease organisms. The immune system of stressed fish is depressed and they are, as a result, more susceptible to disease. When stress is reduced or eliminated fish will perform to their potential. Thus, fish producers must be stress managers.

Stress can be caused by anything that poses a threat to the fish. The response of fish to stress is to devote energy or other resources to help combat the stress. This stress response is adaptive as long as it is short term. When fish are stressed by something like chemical treatments, handling, seining, poor water quality, etc., they devote a lot of energy to fighting that stress. In the short term this is good; it is a beneficial adaptive behavior. However, long-term exposure to stress is harmful.

Long term exposure is a relative term and, depending upon the stress and its severity, may be only a few hours. The energies of the fish are devoted to fighting the stress, but these same energies are also essential for normal balance of the fish. Under chronic stress situations, the fish devotes so much energy trying to stay alive and maintain an equilibrium that the energy is not going to muscle tissue or to maintaining the immune system. Under chronic stress, fish will not grow as well; therefore feed is being wasted and the fish is not gaining in size or weight. Additionally, the immune system will not work as well, so the fish will be even more susceptible to disease. The better these stressful events are managed, or even prevented from occurring, the better the fish are going to do, the better feed conversions are going to be and the less disease will be seen.

Good stress managers have higher quality products and higher profits since less drugs and chemicals are used.

DISEASES AND TREATMENT

It is important to observe fish daily to see the first sign of a disease. Signs that fish may be getting sick fall into two main categories:

- **Behavior - The way fish act, particularly a reduction in feeding activity, will often indicate the beginning of a disease.**
- **Physical appearance - Check any abnormalities to see if a disease is starting.**

Anything that prevents stress will reduce the chance of an infectious disease. Environmental stress is the most common and considering the following pointers can minimize it:

Checking oxygen levels in ponds

- **Oxygen - Should be maintained at a minimum of 4 parts per million for most species, but the maximum should not exceed 150 percent of saturation for 4-6 hours.**
- **Temperature - Should range from 33°F (0.6°C) to 120°F (48.9°C), depending upon the species. Don't exceed a rapid 10°F (5.6°C) increase or decrease or thermal shock is likely in most species.**
- **pH - Most species can survive in pH from 4.5 to 10.5; best range is 6.5 to 9.0.**
- **Nutrients - feed must contain all essential nutrients.**
- **Proper handling.**
- **Prevent toxins, natural or introduced, from entering pond.**
- **Water Quality - Maintain good water quality by preventing buildup of nitrite, ammonia, etc.**

Before treating fish for disease there are several things you need to know:

- **Prognosis.**
- **Feasibility of treating in facility where fish are located.**
- **Economics of treating.**
- **Does loss rate warrant treatment?**

A professional diagnostic lab can provide assistance in dealing with fish-health problems

Before deciding on treatment to use, you must know these things:

- **Disease and how it will respond to the treatment selected.**
- **Water and how it will affect the treatment.**
- **Fish and how they will respond to the treatment.**
- **A specific drug and its effectiveness in a particular situation.**

Drugs and Chemicals Used in Fish Production

The unique, controlled environment in which farm-raised fish are produced results in few human health concerns. Careful site selection for ponds, good water quality management, short production times and rapid processing means that there is only a limited possibility for exposure to chemical or microbial contaminants.

Although farm-raised fish are not likely to be impacted by the environmental pollutants or contaminants that may pose safety concerns in fish captured from the wild, there is one safety concern that is unique to farm-raised fish. This is related to the use of feed additives, drugs and other chemicals. The potential for drug or other chemical residues in farm-raised fish is low, however, when producers follow established federal guidelines and drug and chemical label instructions. Currently there are very few drugs and chemicals that are approved for aquaculture.

Preventing problems before they occur is good management. Management relies on prevention rather than drugs. However, all managers at one time or another must rely on medicated feed or chemicals to treat infectious diseases in their fish. These drugs are used mainly for two reasons: to prevent disease, or to treat disease and relieve animal suffering. By following appropriate Quality Assurance guidelines, farm-raised fish producers ensure that their product is of the highest safety and quality.

Labeling of Fish Health Products

Product information considered essential to ensure the safe and effective use of drugs and chemicals in aquaculture is made available on product labeling. By law, product labels must contain information on feeding rate, dosage to apply, withdrawal periods, proper usage, storage, disposal and any other constraints on the use of the product, including environmental or human safety precautions related to its use.

It is important for fish farmers to read and understand a product label before using the product.

The label will usually describe the following:

1. **Conditions under which the product can be expected to be effective and safe,**
2. **Usage of the product allowed by law,**
3. **Proper storage and disposal of compounds and their containers,**
4. **Precautionary statements on possible environ mental, physical and/or chemical hazards.**

Aquatic Animal Drugs

The U.S. Food and Drug Administration (FDA) oversees the use of drugs in industries that produce animals for food; therefore drug use in fish farming and other types of aquaculture comes under FDA regulation. The list of approved drugs and chemicals for fish is short, and constantly changing. Producers should always check current labels and approvals before applying any material.

Responsibility of Handling and Application of Medication

Control of diseases with drugs should be carried out only on the basis of an accurate diagnosis by a fish disease specialist. An accurate diagnosis is essential for proper selection and usage of drugs. The person making the diagnosis will often use the services of a diagnostic laboratory for fish diseases. Laboratory examination of fish is often essential for proper drug selection and for ensuring proper drug usage. The producer is responsible for maintaining a general health maintenance liaison with fish health specialists. All persons feeding medicated feeds or applying drugs and chemicals must be adequately trained in the proper use of the drug or chemical.

When drugs used in aquaculture are handled or applied it is important to recognize that potentially hazardous effects may occur in fish, in human operators and/or the environment. Producers should communicate closely with fish health specialists regarding dosage and methods of use, and calculation of withdrawal periods.

Uncontrolled and unlimited use of drugs or chemicals could lead to the accumulation of undesirable residues in the fish treated, as well as in the environment. Additionally, continuous use of medicated products may favor the development of disease organisms that are resistant to the drugs or chemicals being used. Fish producers understand and appreciate the importance of sound management and good husbandry in order to reduce the likelihood of fish disease. Every effort should be made to use only those approved medications known to be effective in treating the specific disease.

Extra-Label Use

Disease circumstances sometimes occur where no approved product exists for treatment, or fish applications are not provided for in the product literature or label. In these cases, a licensed veterinarian can, on his/her own responsibility, and under narrowly defined criteria, prescribe or use other licensed products. Such use is called "extra-label" use. Extra-label use by persons other than veterinarians is illegal except when such use is conducted or permitted under the supervision or prescription of a veterinarian. Since few fish health specialists are also veterinarians, extra-label use of drugs and chemicals in fish production is rare.

Extra-label use of products may have unpredictable side effects and have the potential to give rise to unacceptable residue levels. Veterinarian fish disease specialists only embark on such uses after the most deliberate consideration of the needs of the disease situation. Under these circumstances, a significantly extended withdrawal time is assigned for drug withdrawal prior to marketing the fish. The veterinarian is responsible for providing written instructions to the producer on the use and withdrawal times for all extra-label drugs or chemicals.

Extra-label use of drugs applied in the feed is currently not permitted by the FDA.

Withdrawal Times

Federally approved medications that may cause a residue have a "withdrawal time" specified on the label. Withdrawal information is available on the product label, package insert or the feed tag. Withdrawal times for fish medications have been established to ensure that safe, wholesome fish products reach consumers. Withdrawal time varies depending on the drugs involved, but usually is listed as a specified number of days from the last application of the drug.

To avoid the presence of unacceptable residues in fish or by-products of fish origin, it is essential that fish producers adhere to the withdrawal time established for each product and dose regime.

For example, when using a drug with a listed withdrawal time of 7 days, a producer must record the last application date and time. He then must wait a full 7 days (1 day is defined as 24 hours) before any of the treated fish can legally be harvested and marketed for human consumption.

Anyone who adds drugs to aquatic feeds, whether a feed manufacturer or a fish producer, is held responsible for proper use, including correct dosage and observation of withdrawal times.

All drugs must be approved and shown to be safe and effective, with clearly established withdrawal times. It is very important that medications be added at the levels specified by the label. If not, a violation occurs.

Environmental Considerations

Fish producers and fish disease specialists recognize and respect the fact that any use of drugs or chemicals will have an influence on the aquatic environment.

Fish treated will normally include both healthy and sick individuals. The aquatic environment is then exposed to relatively large amounts of medication through feces and uneaten medicated feed.

To avoid their accumulation in the environment, it is desirable to limit the use of fish health medications to the lowest possible levels. Proper use is also important to protect the surrounding environment, which can be harmed when products are used incorrectly. Treated waters can potentially carry medication to distant locations where treatments may affect non-target organisms and sites.

Record Keeping

Records are essential for any business, and aquaculture is no exception. Record keeping is the key to making good management decisions. Records can also be a protective measure against potential liability. Most importantly, records are a necessary component of a producer's quality assurance program. Good records detail withdrawal status of treated ponds of fish.

Fish producers should keep records of the products used, including the quantity, the date and frequency of application, time of harvest and the identity of fish (or batches of fish) on which the medicines were used. Records should be kept for at least two years. When possible, include written prescriptions from veterinarians in records

Storage and Disposal of Fish Medications and Chemicals

Fish producers should always read and follow label directions for storage and disposal of fish drugs and chemicals.

1. **Store all fish health products in a secure area, under lock and key where practicable.**
2. **Keep all drugs and chemicals out of reach of children and animals.**
3. **Properly dispose of drugs remaining after treatment has been completed.**
4. **Never retain empty or partially used containers for future use.**
5. **Return unused drugs that are beyond their expiration date to the vendor if there is an agreement to that effect.**
6. **Advise users about correct disposal measures to reduce potential contamination of the environment.**

Pesticides

Certain groups of pesticides, such as herbicides, algaecides, insecticides, limacides and piscicides can be useful in aquaculture. The U.S. Environmental Protection Agency (EPA) is responsible for regulating pesticide use on fish farms. State agencies are also often involved in pesticide regulation. Fish producers should apply only those compounds that are registered by EPA and their state for use in aquaculture on food-fish.

Application of pesticides to culture facilities and their surroundings must be done with great care and with adherence to the instructions given by the manufacturer or the specialist. Persons carrying out the application of pesticides must be adequately trained to understand the application procedure and the risks. Training and certification are required for use of any restricted use pesticide. Private or commercial pesticide applicator training and certification procedures may be obtained by contacting your county extension agent.

Aerial applicators are supplied with guidelines for applying pesticides near fish ponds. Producers and applicators work together to ensure that necessary pesticide applications on neighboring farms do not compromise the safety and quality of farm-raised fish. Agricultural pesticides are categorized as to their potential for drift and their relative toxicity to fish. Pesticides with high drift potential and/or high toxicity are not applied near fish ponds.

12 Harvesting

The method you use for harvesting your crop will depend on your production system. Tanks, raceways and other small, contained units are the simplest to harvest, whereas large, multi-acre ponds pose the greatest logistical challenges.

Some producers, especially those with large amounts of fish or large production units, choose to hire a custom harvest crew to harvest the crop. Custom harvesters usually supply most of their own equipment, and charge a fee to harvest the animals. If you plan to employ a custom harvester, a few simple guidelines will be important to remember:

1. **Agree on the fees and costs BEFORE the harvest.**
2. **Arrange for the harvest in advance, so that both you and the harvester are adequately prepared.**
3. **Discuss with the harvester what equipment he will supply and what equipment he expects you to supply. For example, on pond systems, the harvester usually expects the farmer to provide one or more tractors, an aerator, and sometimes other equipment and supplies.**
4. **Be sure that you have a market for your fish before any product goes into the pond.**

Seine harvesting prawns in a small scale operation

Some producers with ponds, and most with small-scale tanks, raceways, etc., prefer to do their own harvesting. Thus, certain equipment will be required, depending on the kind of system you use. Harvesting from cages is simply a matter of using dip nets to lift fish from cages to be placed in transport containers. Harvesting from ponds, however, may require tractors, large seines, holding sacks on the seines, boats, waders, a boom and basket for lifting fish from the pond onto transport trucks.

Seine in position during a prawn harvest

Any harvest operation will require the use of some means of weighing or counting fish as they come out of the pond, depending upon the species and the market demands. Some species must be immediately chilled with ice or ice-water upon removal from the culture system, while others are transported live to a processing or other distribution point. Be sure that you are prepared with all necessary equipment and supplies to allow you to adequately handle, move and then market the crop that you remove from your production facility.

Crawfish farmers use boats and baited traps to harvest their product

An important consideration in any harvest operation is the marketability of the product. It is important that you ensure that your market is in place and ready to receive the product BEFORE harvest! In species where off-flavor is a marketing constraint, it is vital that the buyer, whether processor or other distributor, approve of the sample of fish PRIOR to harvest. Many producers have gone to the trouble and expense of harvesting and transporting fish only to find that they had not adequately ensured their market. When this happens, a substantial loss can be incurred, either through stress on the fish from the handling and transport, or from reduced marketability of the product.

13

Marketing Your Product

Perhaps the most important consideration in planning your aquaculture operation is the one most often overlooked—marketing. In any aquaculture business, the bottom line to success is being able to sell your product at a PROFIT! Thus, one of the first lines of thought should be identification and establishment of market channels for your product. A good marketing strategy will pay economic dividends in the future once your production facility is in place and producing a crop.

A good marketing strategy consists of three basic elements:

1. **Analysis of present market conditions and opportunities**
2. **Development of specific goals to develop and secure a place in the market**
3. **Development of a logical and orderly plan for attaining your specific marketing goals.**

Analysis of Present Market Conditions

Step #1 in your marketing strategy involves doing a significant amount of "homework". It will involve talking to market contacts, assessing the market from several perspectives, assessing your ability or potential to meet particular market demands, and assessing your potential production system and its capacity, potential, limitations and strengths relative to providing a marketable product in adequate quantity that is also of acceptable quality. Only after assessing current conditions and opportunities can you feasibly develop your specific goals.

Developing Specific Marketing Goals

Once you have made appropriate assessments of current markets, you can set reasonable goals for tapping the desired market(s). Matching your potential with a recognized and available market niche is the key to marketing success. You may identify many market opportunities that, for one reason or another, you cannot supply on a consistent, dependable basis. Search until you find a comfortable match between your production capabilities and a market niche that exists. From that point, it will be simply a matter of implementing a plan to attain the established goals.

Develop a Plan

Your marketing plan will involve consideration of many aspects of your production operation and the existing market. You should consider the costs of production, your production efficiencies and production capacities. Once these have been matched with a market outlet that you can satisfy, establish the procedures you will use to access and supply that market. Consider contingency plans in case of market changes, production shortfalls or other unfortunate occurrences.

Once the plan is finalized, write it down and review it often. The marketing world is dynamic, and changes in the marketplace must be monitored. Your production strategy, timing for harvests, and many other logistical concerns will be dependent upon the marketing strategy you develop. If, and when, that marketing strategy must change so, too, must your production strategy.

Processor's truck picking up fish at the pond bank

Marketing Options

The number of specific marketing options for aquaculture products could be almost infinite, but most will fit into one of the following broad categories.

Processing Plants

For many producers, this is the easiest, most convenient, and most efficient marketing outlet. Processing plants do not exist for many species, however, and it is important to make this determination before deciding upon a species for culture. Most plants will prefer to buy the product in bulk, and some, such as in the catfish industry, will buy the fish on the pond bank, then transport the product to central processing facilities. Other processors may require that you transport the product to their plant, which becomes the point of sale.

Most processors will require advance notice of product availability, and will often have strict quality and taste standards to which the product must adhere. Some plants will require preharvest quality and flavor testing of the product before committing to the purchase.

Live Haulers

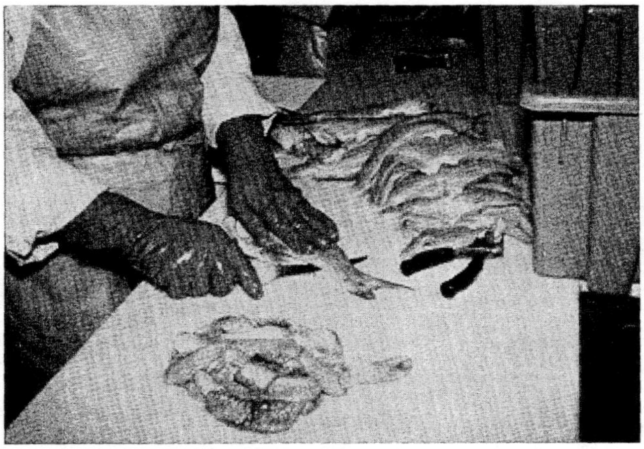

Catfish moving through a processing plant

Live haulers purchase the product usually at the production site and then transport to distant markets. In this sense, live haulers function as intermediate brokers for aquaculture products. For some species, live hauler sales is the primary market opportunity, while live-hauling does not exist for other species.

Many live haulers transport fish to be processed at another location, or to be marketed either live or without further processing. Live haulers are important to small niche markets, since these markets and their consumers often exist far away from the actual production facilities for a particular species.

Other live haulers transport fish to re-stock into ponds or other units, typically for marketing through "fish-out" or "pay-lake" operations, which are popular in many parts of the country.

A live-hauler's truck at the pond bank

Local Restaurants and Stores

Many small-scale producers are successful in identifying and securing a local market for their product. Amenable arrangements can often be made with local vendors or restaurants, and these market outlets usually have a limited, but consistent need for product. Such markets, however, often require processed or partially processed product, and you may be required to set up a small-scale processing plant to supply this kind of market.

Direct Sales

Many species can be effectively marketed on a local basis direct from the pond or production facility. This is most effective with specialty products that are produced on a limited, small-scale basis. Success of this strategy will depend on geographic location, proximity to population centers, the size of you operation and daily or weekly yield, and the alternative availability of the product from other market channels.

Fee-Fishing

Fee fishing is a popular marketing option

Some producers, especially of species that are sought by anglers, can establish a marketing outlet by employing a "fee-fishing" strategy. Fee-fishing involves allowing access by anglers to one or more designated ponds and charging a fee either for access to fishing privileges or charging on a per unit basis for the fish that are caught by hook-and-line.

Fee fishing ponds are sometimes called catch-out ponds. They usually are small and heavily stocked with one or more types of fish of catchable size. There are three basic types of fee fishing operations: long-term leasing, day leasing, and fish-out.

Long-term leasing involves providing fishing privileges at a pond or lake to an individual or to a group. The length of the agreement and the specific services to be provided are spelled out in the lease.

Day leasing is similar to long-term leasing, but the pond or lake is only leased on a half-day or daily basis. The operator of the fish-out pond generally charges a basic fee for one-half or one day of fishing. There also may be an additional charge for each pound or inch of fish caught.

Fish-out pond operators usually charge their customers a set fee for each pound or inch of fish caught. In addition to the pond or lake, operators also may provide a snack bar, bait and tackle shop, picnic facilities, rest rooms, parking areas and boat rentals.

Common Calculations Used in Fish Pond and Lake Management

14

Drugs and pesticides are applied to ponds and tanks for many reasons, including:

1. **To control fish diseases.**
2. **To control aquatic weeds.**
3. **To selectively eliminate specific kinds of undesirable fish.**
4. **To correct undesirable water quality problems.**

For several good reasons, it is absolutely essential that the amount of material to use for a particular problem be accurately calculated. If you overtreat, there is a high probability that some, if not all, of the fish will be killed. If you undertreat, you won't get the desired effect and, in the case of a disease, all fish may die before you can re-treat the pond or tank. In either case, a miscalculation can be expensive.

You must know the correct volume of water to be treated before attempting to calculate the amount of material to use. To obtain the volume of a pond, determine the surface acreage and the average depth of the pond. Then multiply the surface acres x the average depth to get the volume of the pond in acre-feet.

To determine the volume of a tank, raceway, or holding vat, measure its width and length and the depth of water to be present at the time of treatment. Multiply these three measurements together to calculate the volume. The measurements can be in inches, feet, yards, centimeters, meters. In other words, the unit of measurement used is not important, although measuring in feet is usually most convenient.

After determining the correct volume of water to be treated, use the formula to calculate the correct amount of material to use.

Amount of material needed = Vol. x Conv. factor x ppm desired x (100 / % active ingredient).

Vol. = volume of water to be treated. Can be any volumetric units of measurement - cubic feet, liters, gallons, acre-feet.

Conversion factor = the weight of material that must be added to one unit volume of water to give one part per million (ppm). See Table 1 for conversion factors.

ppm desired = the treatment rate desired in parts per million (ppm).

% active ingredient = the percent ingredient of the material to be used. Most materials used in treating fish diseases or aquatic weeds are 100% active or are considered to be 100% active for treatment purposes.

Table 1. Weight of chemical that must be added to one unit volume of water to give one part per million (ppm) (conversion factors)

2.72 pounds per acre-foot = 1 ppm

1,233 grams per acre-foot = 1 ppm

0.0283 grams per cubic foot = 1 ppm

0.0000624 pounds per cubic foot = 1 ppm

0.0038 grams per gallon = 1 ppm

0.0584 grains per gallon = 1 ppm

1 milligram per liter = 1 ppm

0.001 gram per liter = 1 ppm

8.34 pounds per million gallons of water = 1 ppm

Table 2. Conversions for units of volume

To From	cm3	liter	m3	in3	ft3	fl oz	fl pt	fl qt	gal
cm3	1	0.001	1×10^{-6}	0.0610	3.53×10^{-5}	0.0338	0.00211	0.00106	2.64×10^{-4}
liter	1000	1	0.001	60.98	0.0353	33.81	2.113	1.057	0.2642
m3	1 x 106	1000	1	6.1×10^{4}	35.31	3.38×10^{4}	2113	1057	264.2
in3	16.39	0.0164	1.64×10^{-5}	1	5.79×10^{-4}	0.5541	0.0346	0.0173	0.0043
ft3	2.83×10^{4}	28.32	0.0283	1728	1	957.5	59.84	29.92	7.481
fl oz	29.57	0.0296	2.96×10^{-5}	1.805	0.00104	1	0.0625	0.0313	0.0078
fl pt	473.2	0.4732	4.73×10^{-4}	28.88	0.0167	16	1	0.5000	0.1250
fl qt	946.4	0.9463	9.46×10^{-4}	57.75	0.0334	32	2	1	0.2500
gal	3785	3.785	0.0038	231.0	0.1337	128	8	4	1

Table 3. Conversions for units of length

From To	cm	m	in	ft	yd
cm	1	0.01	0.3937	0.0328	0.0109
m	100	1	39.37	3.281	1.0936
in	2.540	0.0254	1	0.0833	0.00278
ft	30.48	0.3000	12	1	10.3333
yd	91.44	0.9144	36	3	1

Table 4. Conversions for units of weight

To From	gm	kg	gr	oz	lb
gm	1	0.001	15.43	0.0353	0.0022
kg	1000	1	1.54 x 104	35.27	2.205
gr	0.0648	6.48×10^{-5}	1	0.0023	1.43×10^{-4}
oz	28.35	0.0284	437.5	1	0.0625
lb	453.6	0.4536	7000	16	1

Table 5. Miscellaneous conversion factors

1 acre-foot	equals	43,560 cubic feet
1 acre-foot	equals	325,850 gallons
1 acre-foot of water	equals	271,814 pounds
1 cubic foot of water	equals	62.4 pounds
1 gallon of water	equals	8.34 pounds
1 gallon of water	equals	3785 grams
1 liter of water	equals	1,000 grams
1 fluid ounce	equals	29.57 grams
1 fluid ounce	equals	1.043 ounces

Temperature Conversions

Centigrade to Fahrenheit = (C x 9/5) + 32

Fahrenheit to Centigrade = (F - 32) x 5/9

Abbreviations

cm = centimeters	ft^3 = cubic foot	kg = kilogram
cm^3 = cubic centimeters	gallon = gallon	lb = pound
fl oz = fluid ounce	gm = gram	m = meter
fl pt = fluid pt	gr = grain	m^3 = cubic meter
fl qt = fluid quart	in = inch	oz = ounce
ft = foot	in^3 = cubic inch	yd = yard

15

Aquaculture Resources

One of the best sources of additional information on aquaculture is your local or area Cooperative Extension service (county agent) office. In most states, there is an office in each county, serving the informational and educational needs of the general public. The information available covers a wide range of subjects. Although not every local office has aquaculture information on hand, ALL local extension offices are part of the extension network and can access subject matter specialists and information from the land-grant university or universities in your state and in other states.

In most states where aquaculture exists, there are designated aquaculture, fisheries or sea-grant extension specialists who are familiar with aquaculture literature and the industry, as well as the specific opportunities and potentials in their respective states or territories. These are the best, first-contact sources for information to expand your knowledge of aquaculture.

Hundreds of publications exist (far too many to list here), and your state extension or sea-grant extension office can help obtain copies. Extension agents and specialists also can provide contacts with other local, state and federal, public and private agencies involved in aquaculture.

Another invaluable source of information and communication is the World Wide Web. There are hundreds of sites with aquaculture information. One of the most comprehensive is the Aquaculture Network Information Center (AquaNIC), a national project by aquaculture specialists and scientists.

AquaNIC logo

The Aquaculture Network Information Center

The Aquaculture Network Information Center (AquaNIC) is a gateway to the world's electronic resources in aquaculture. AquaNIC is maintained at Purdue University in West Lafayette, Indiana, and is supported by the Illinois-Indiana Sea Grant Program and Purdue University's Department of Animal Sciences. The AquaNIC site contains links to aquaculture sites at other state Land Grant universities, USDA sites devoted to aquaculture, professional organizations, and other sites with aquaculture information.

AquaNIC can be accessed several ways:

1. WWW: http://ag.ansc.purdue.edu/aquanic/

2. TELNET: www.ansc.purdue.edu

Login and password lynx (lower case)

3. MODEM: 317-496-1440

Data bits 8, Parity None, Stop bits 1

User name - lynx (lower case)

Enter - rlogin www.ansc.purdue.edu

Password - lynx

4. FTP: www.ansc.purdue.edu

Fishy 98

Another helpful resource for anyone engaged in an aquaculture enterprise is Fishy 98, a computer program for fish farmers. The program is designed for farm-raised catfish, but it is adaptable to other fish species, such as trout, tilapia and hybrid striped bass.

Fishy 98 is designed to help farmers make better management decisions, keep up with the history of their ponds, figure the out-of-pocket costs of growing fish and cope with end-of-the-year record keeping.

The program is available to producers at no cost. Additional information is available by contacting the Mississippi State University Department of Agricultural Economics at (601) 325-2750.

Getting Started in Aquaculture

A series of species profiles have been developed by LaDon Swann, Extension Aquaculture Specialist, Illinois-Indiana Sea-Grant College Program, Purdue University. These profiles are part of a multi-media CD-ROM product entitled "Getting Started in Aquaculture," which is an excellent resource and decision making tool for the prospective fish farmer. Among the many other kinds of helpful information, "Getting Started in Aquaculture" contains a wealth of economic, marketing and business planning assistance and documents that will provide helpful insight and guidance to readers of this book once the decision is made to establish a business. The following species profiles are from "Getting Started in Aquaculture" and are used courtesy of LaDon Swann and the Illinois-Indiana Sea-Grant College Program.

Line drawings of fish species reproduced with permission from "How to Know the Freshwater Fishes," by Samual Eddy and James C. Underhill, 3rd edition (1978), McGraw-Hill Publishing Companies, unless otherwise noted.

COMMON AQUACULTURE SPECIES

Atlantic Salmon

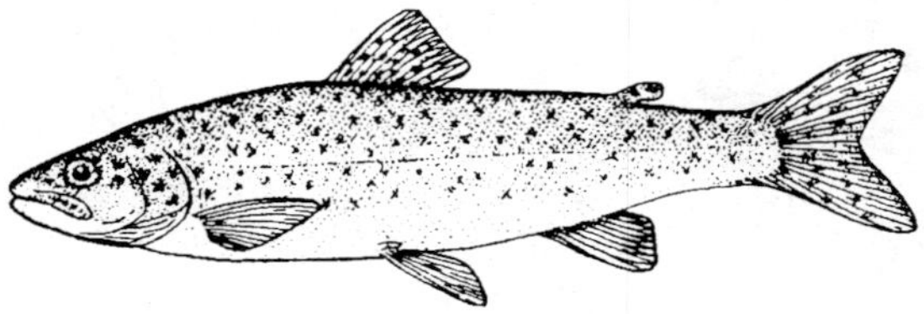

Scientific name: Salmo salar

Production potential: Easy

Marketing potential:

Potential: High

Size: 1.5-5.0 pounds

Market: Food

Temperature requirements:

Growing: 50-62°F

Spawning: 42-50°F

Lethal: NA

Feed requirements:

Protein: 45-50%

Fat: 25%

Spawning requirements: Spring spawners, 800 eggs/lb. body weight

Most common production systems: Raceways and net pens

Black Bullhead

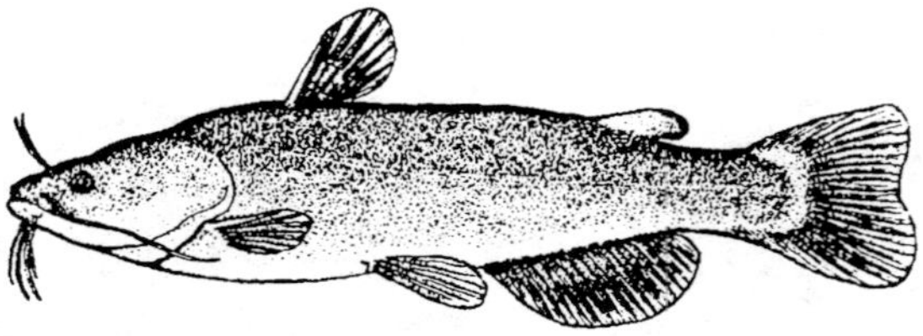

Scientific name: Ictalurus melas

Production potential: Easy

Marketing potential:

Potential: Moderate

Size: 0.75-1.0 lb.

Market: Food and sport

Temperature requirements:

Growing: 78-82°F

Spawning: 80°F

Lethal: NA

Feed requirements:

Protein: 32%

Fat: NA

Spawning requirements: Spawning occurs between mid-May and July. Females build nests and the male guards the nest after the first day. Females may produce 2,500-3,800 eggs per fish. Eggs hatch in 5-10 days.

Most common production systems: Ponds and cages. Ponds stocked at 2,000 fingerlings per acre. Cages stocked at 8-10 fingerlings per cubic foot.

Blue Catfish

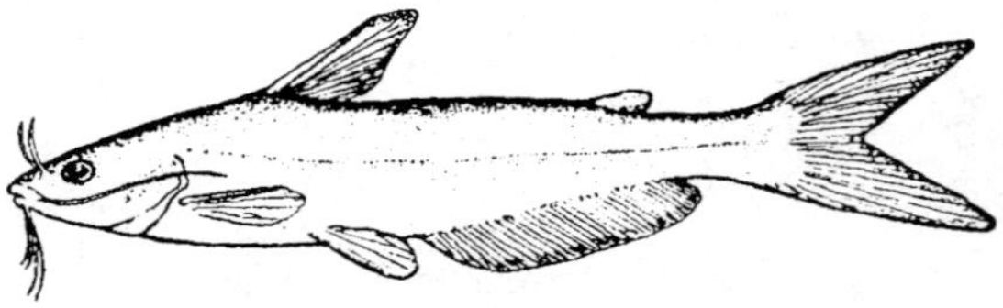

Scientific name: Ictalurus furcatus

Production potential: Moderate

Marketing potential:

Potential: Medium

Size: 0.75-2.5 lb.

Market: Food, sport

Temperature requirements:

Growing: 78-82°F

Spawning: 72-82°F

Lethal: NA

Feed requirements:

Protein: 28-30%

Fat: 6-12%

Spawning requirements: Annual spring spawners that lay eggs in cavities at a rate of 3,000-4,000 eggs/lb body weight. Eggs hatch in 7-8 days at 78°F. There is considerable interest in hybrids produced from channel catfish and blue catfish.

Most common production systems: Ponds and cages. Ponds stocked at a rate of 2,000-4,000, 4-8 inch fingerlings per acre. Cages are stocked at a rate of 6-10 fish per cubic foot.

Bluegill

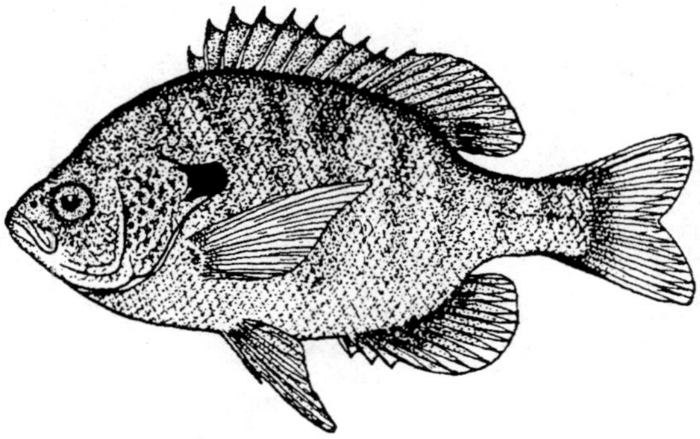

Scientific name: Lepomis macrochirus

Production potential: Easy

Marketing potential:

Potential: Moderate

Size: Three fish per pound

Market: Sport, food

Temperature requirements:

Growing: 55-80°F

Spawning: 75-80°F, best at 80°F.

Lethal: NA

Feed requirements:

Protein: 32%

Fat: 8-10%

Spawning requirements: Spring spawners, nest builders, 50,000 eggs/lb body weight

Most common production systems: Ponds

Channel Catfish

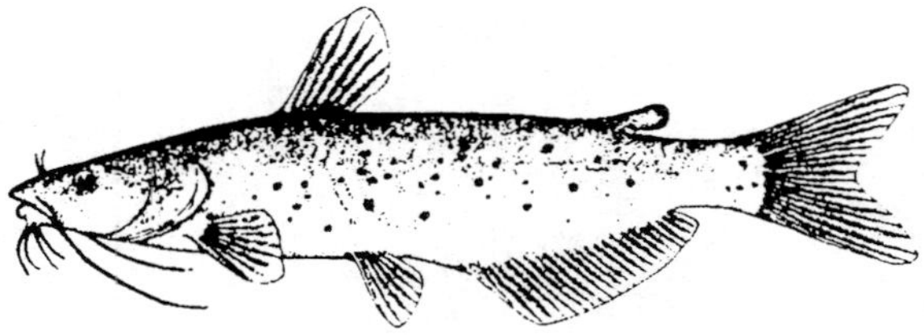

Scientific name: Ictalurus punctatus

Production potential: Easy

Marketing potential:

Potential: High

Size: 0.75-2.5 lb.

Market: Food, sport

Temperature requirements:

Growing: 80-85°F

Spawning: 72-82°F

Lethal: NA

Feed requirements:

Protein: 28-30%

Fat: 6-12%

Spawning requirements: Annual spring spawners that lay eggs in cavities at a rate of 3,000-4,000 eggs/lb body weight. Eggs hatch in 7-8 days at 78°F.

Most common production systems: Ponds and cages. Ponds stocked at a rate of 2,000-4,000, 4-8 inch fingerlings per acre. Cages are stocked at a rate of 6-10 fish per cubic foot.

Coho Salmon

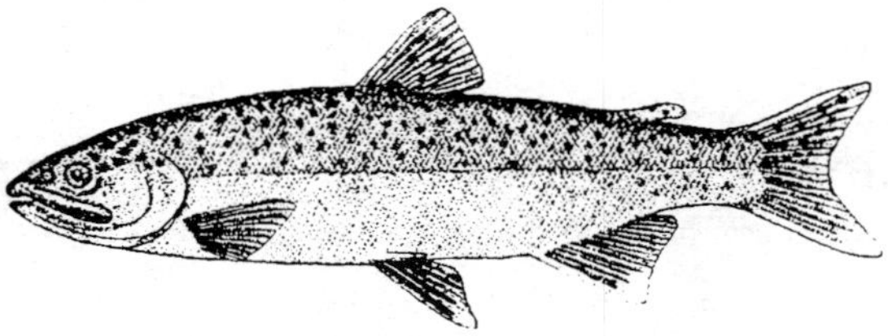

Scientific name: Oncorhynchus kisutch

Production potential: Easy

Marketing potential:

Potential: High

Size: 0.5-10.0 pounds

Market: Food

Temperature requirements:

Growing: 48-58°F

Spawning: 45-55°F

Lethal: NA

Feed requirements:

Protein: 45-50%

Fat: 16-18%

Spawning requirements: Spawn once in life then die; 400 eggs/lb body weight

Most common production systems: Raceways and pens

Common Carp

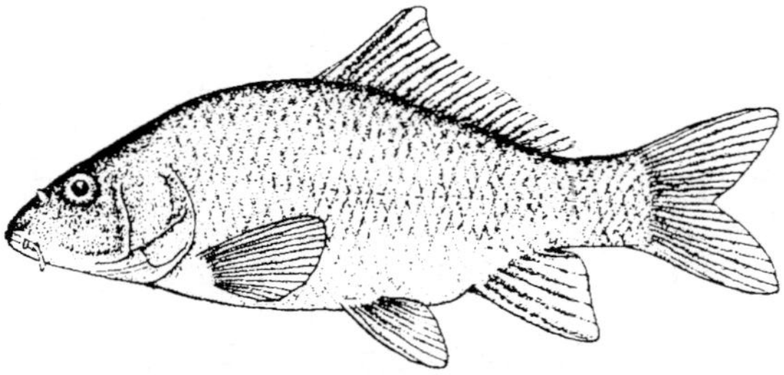

Scientific name: Cyprinus carpio

Production potential: Easy

Marketing potential:

Potential: Moderate

Size: 3-5 pounds

Market: Food

Temperature requirements:

Growing: 55-80°F

Spawning: Above 65°F

Lethal: NA

Feed requirements:

Protein: 31-38%

Fat: 3-8%

Spawning requirements: 60,000 eggs/lb. body weight. Eggs hatch in 2-7 days

Most common production systems: Ponds

Fathead Minnow (tuffy)

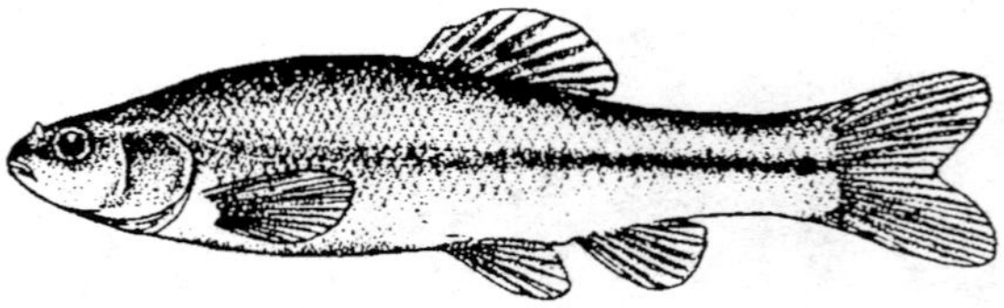

Scientific name: Pimephales promelas

Production potential: Moderate

Marketing potential:

Potential: High

Size: 1-3 inches

Market: Bait

Temperature requirements:

Growing: 70°F

Spawning: 65°F

Lethal: NA

Feed requirements:

Protein: 30-38%

Fat: 3-8%

Spawning requirements: Spawns during cool periods in summer and sometimes in fall, eggs hatch in 5-6 days. Females lay from 200-500 eggs per spawn, and spawn repeatedly during spawning season. Two methods are used to produce fry. The first is the wild or free-spawning method, which 500-2,000 broodstock are stocked into ponds. Egg hatching and growing of young to marketable size occur in the same pond. The second method is the fry transfer method. In this method, 20,000-25,000 broodstock are stocked into ponds in a ratio of five females per male. Fry produced in either method are trapped and counted, then transferred to rearing ponds. In either method, rocks, tile, bricks, or boards are placed into ponds to provide spawning sites.

Most common production systems: Ponds. Fry are stocked at a rate of 50,000-300,000 per acre.

Freshwater Shrimp

Photo by Bob Ratliff

Scientific name: Macrobrachium rosenbergii

Production potential: Moderate

Marketing potential:

Potential: High

Size: 6-10 per pound

Market: Food

Temperature requirements:

Growing: 83-85°F

Spawning: NA

Lethal: Lower 50-55°F

Feed requirements:

Protein: 35-40%

Fat: NA

Spawning requirements: Around 12 parts per thousand (ppt) salinity is required for spawning. Females spawn in freshwater and lay 10,000-50,000 eggs per spawn, 4-5 times per year when water temperatures are above 69°F. Eggs are attached to the pleopods of females 6-20 hr. following mating. Eggs hatch in approximately 19 days when the water temperature is above 82°F. Newly hatched larvae must return to salt water where they will undergo 11 molts in approximately 35 days. At this stage, they are postlarvae and are 7-10 mm in length and resemble adults. Postlarvae are cannibalistic and omnivorous. Once the postlarvae reach 0.1-0.3 grams, they are at the juvenile stage with a body color of blue to brown. The time to a juvenile stage is from 60-90 days. Adult shrimp are marketable at 2-3 ounces (40-70 grams).

Most common production systems: Ponds. Juveniles grow best at 84-88°F, but will survive temperatures between 57-95°F. Juveniles can be stocked in the spring when water temperatures reach 68°F at a rate of 40,000 per hectare (16,187 per acre). Ponds should be fertilized to obtain a zooplankton bloom; and once they reach 5 grams, may be fed a sinking 32% fish feed. Prawns are harvested when water temperatures drop to 68°F. Survival varies widely, but should be between 50-90%.

Golden Shiner

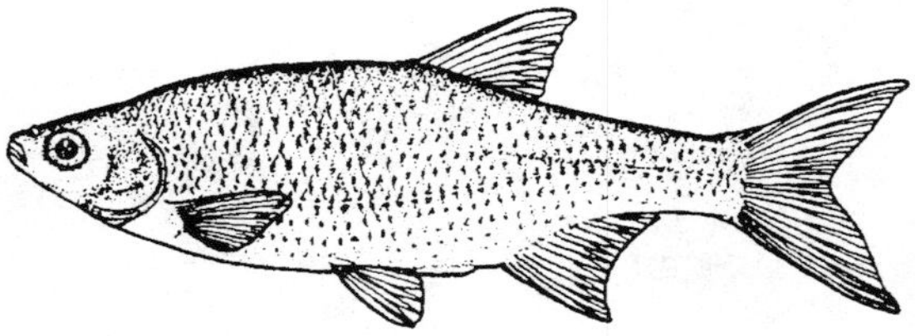

Scientific name: Notemigonus crysoleucas

Production potential: Easy

Marketing potential:

Potential: High

Size: 2-6 inches

Market: Bait

Temperature requirements:

Growing: 50-80°F

Spawning: 70-85°F

Lethal: NA

Feed requirements:

Protein: 30-38%

Fat: 3-8%

Spawning requirements: Spawns May-June, 75,000 eggs/lb body weight. Eggs hatch in 4-9 days at spawning temperature. Three methods are used to produce fry. The first is the wild or free-spawning method which 20-40 pounds of broodstock are left to spawn in ponds. Egg hatching and growing of young to marketable size occur in the same pond. The second method is the egg transfer method. In this method the broodstock spawn on spawning mats placed in shallow water along the shore. When mats are covered with eggs, they are moved to rearing ponds. The final method is the fry transfer method. In this method, fry produced from either wild spawning or the egg transfer method are trapped and counted then transferred to rearing ponds.

Most common production systems: Ponds. Fry are stocked at a rate of 50,000-300,000 per acre.

Goldfish

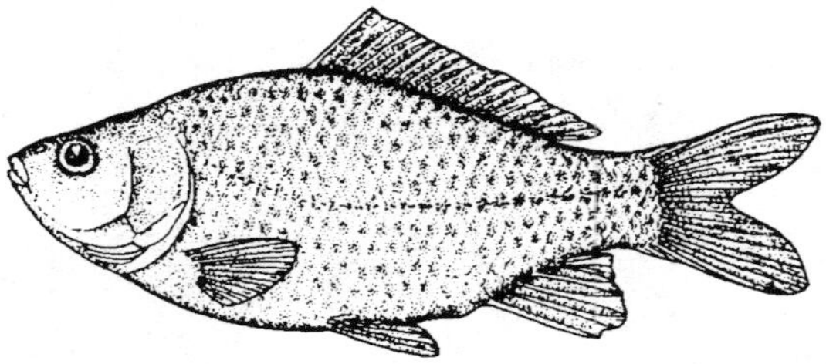

Common name: The comet variety is the most common type of goldfish, but there have been many other varieties developed such as black moors, calico, koi, and shubunkins.

Scientific name: Carassius auratus

Production potential: Easy

Marketing potential:

Potential: Moderate

Size: 1-6 inches for ornamental and 1-2 inches for feeder fish.

Market: Ornamental

Temperature Requirements:

Growing: 70°F

Spawning: Above 60°F

Lethal: NA

Feed requirements:

Protein: 30-38%

Fat: NA

Spawning requirements: Spawns repeatedly from May to June, eggs hatch in 2-8 days, 50,000 eggs/lb. body weight. The primary method used is the egg transfer method. In this method the broodstock spawn on spawning mats placed in shallow water along the shore. When mats are covered with eggs, they are moved to rearing ponds.

Most common production systems: Ponds. Small ponds 0.25-1.0 acres for spawning and larger ponds 0.5-5 acres for rearing of fry.

Grass Carp

Photo by Martin W. Brunson

Scientific name: Ctenopharyngodon idella

Production potential: Difficult

Marketing potential:

Potential: Moderate - High

Size: 0.5 - 5.0 lb.

Market: Aquatic weed control, food

Temperature requirements:

Growing: 70-77°F

Spawning: 68-80°F

Lethal: NA

Feed requirements:

Protein: 41-43%

Fat: NA

Spawning requirements: Fish become sexually mature in 2-3 years in the south and 4-6 years in the north. In the wild grass, carp are riverine spawners and spawn in the spring when temperatures reach 68-72°F. In aquaculture, grass carp are spawned in tanks or hand stripped and the fertilized eggs incubated in McDonnell jars. Spawning is induced using carp pituitary or gonadothropins. Mature females may produce 500,000-800,000 eggs that can hatch in 18-28 hours.

Most common production systems: Ponds. At 120,000 fry per acre, it will take approximately one month to reach 2 inches. Stocked at rate of 35,000 fry per acre, it will take 130-150 days to reach 5 inches.

Hybrid Striped Bass, Sunshine Bass, Palmetto Bass, Wipper

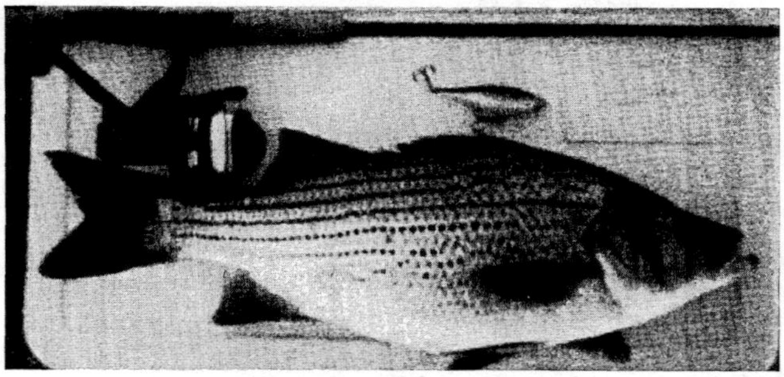

Photo by LaDon Swann

Scientific name: Morone saxatilis x M. chrysops or M. chrysops x M. saxatilis

Production potential: Moderate

Marketing potential:

Potential: Moderate

Size: 1.5-3.5 lbs

Market: Food, sport

Temperature Requirements:

Growing: 77-86° F

Spawning: 61-68° F

Lethal: N.A.

Feed requirements:

Protein: 45-50%

Fat: 15% Fry: Stocked into ponds containing rodifer and cladocern zooplankton populations.

Spawning requirements: Spawns late March to May. Cross made with either female white bass and male striped bass (sunshine bass), or with female striped bass and male white bass (palmetto bass). Female striped bass average 100,000 eggs/lb. body weight (25,000 eggs per ounce of spawn) and white bass females average 50,000 eggs/lb. body weight (100,000 eggs per spawn). Eggs hatch in two to five days. Survival of fry is low, less than 50%.

Most common production systems: Ponds, recycle systems and cages

Hybrid Sunfish

Photo by Martin W. Brunson

Common name: Commonly the cross between the female green sunfish and the male bluegill

Scientific name: Lepomis cyanellus x Lepomis macrochirus

Production potential: Easy

Marketing potential:

Potential: Moderate

Size: Three fish per pound for food. Hybrid sunfish are also sold as fingerlings for stocking farm ponds at 1-3 inches.

Market: Sport, food

Temperature requirements:

Growing: 55-80°F

Spawning: 75-80°F, best at 80°F.

Lethal: NA

Feed requirements:

Protein: 32%

Fat: 8-12%

Spawning requirements: Spring spawners, nest builders, 50,000 eggs/lb body weight

Most common production systems: Ponds

Largemouth Bass

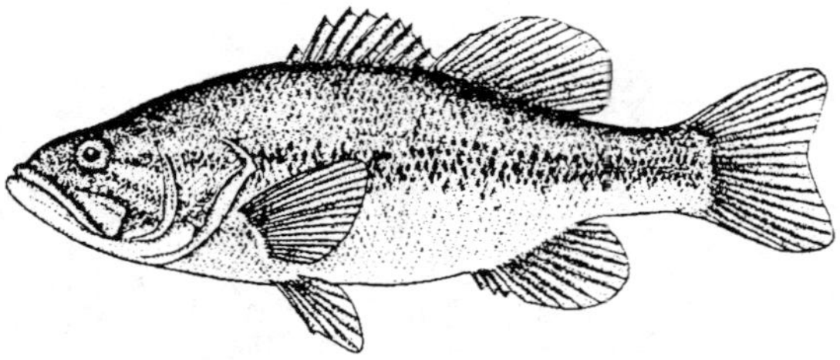

Scientific name: Micropterus salmoides

Production potential: Moderate

Marketing potential:

Potential: High

Size: 1.0-2.5 lb.

Market: Sport, food

Temperature requirements:

Growing: 55-80°F

Spawning: 60-65°F

Lethal: NA

Feed requirements:

Protein: 40%. Forty percent diets are normally fed from fingerlings to adults after fingerlings have been trained to accept commercial diets.

Fat: NA

Spawning requirements: Spawning runs from spring through summer when temperature is 15°F (24°C). Nest layers, 13,000 eggs/lb body weight. Eggs hatch in approximately four days.

Most common production systems: Ponds for spawning and grow-out to food-fish size. Small fingerlings are normally removed from ponds and trained to accept commercial diets using flow-through systems.

Northern Crayfish

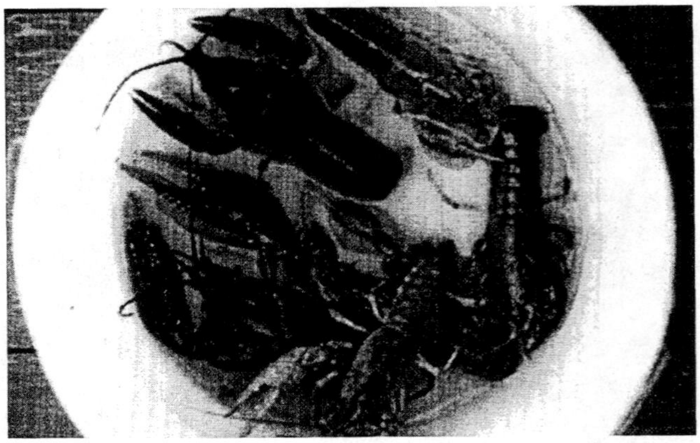

Photo: AquaNIC

Scientific name: Orconectes virilis

Production potential: Easy

Marketing potential:

Potential: Moderate

Size: 7-13 per pound

Market: Bait, food

Temperature requirements:

Growing: 75-85°F

Spawning: 60-70°F

Lethal: 95°F or more

Feed requirements:

Protein: 30%

Fat: 4-6%

Spawning requirements: Northern crayfish spawn March-May. Females lay from 200-500 eggs, which hatch in three to ten days.

Most common production systems: Ponds are stocked with 60 lb of broodfish per acre in the spring. They reproduce during the spring. Crayfish are large enough to harvest in the fall.

Rainbow Trout

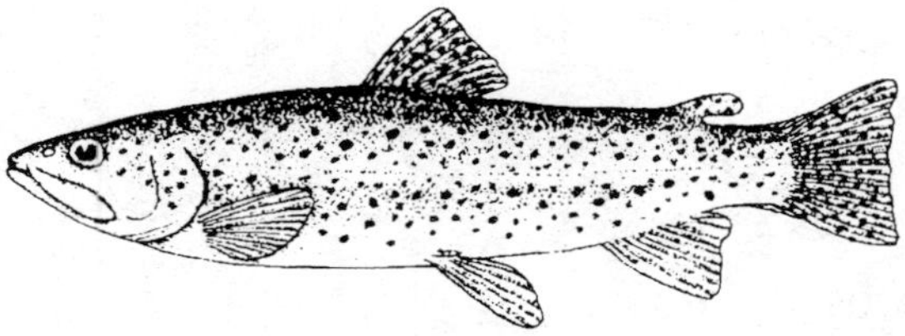

Scientific name: Oncorhynchus mykiss

Production potential: Easy

Marketing potential:

Potential: High

Size: 0.5-1.5 lb.

Market: Food, sport

Temperature requirements:

Growing: 50-60°F

Spawning: 50-55°F

Lethal: Greater than 70°F

Feed requirements:

Protein: 45-50%

Fat: 12-15%

Spawning requirements: Spring and fall spawning strains, 500-10,000 eggs depending on size of female, eggs hatch in 24-31days at 50-55°F.

Most common production systems: Raceways and net pens

Red Drum

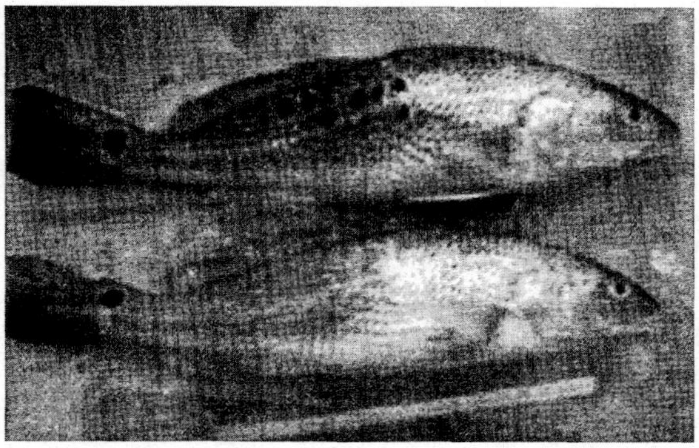

Photo by Jimmy Avery

Scientific name: Sciaenops ocellatus

Production potential: Moderate in southern climates

Marketing potential:

Potential: High

Size: 1.5-2.5 lb.

Market: Food

Temperature requirements:

Growing: 65-85°F

Spawning: 72-75°F

Lethal: 46°F

Feed requirements:

Protein: 35% containing at least 12% fish meal

Fat: 8%

Spawning requirements: Spawning occurs in September through October. Females may produce 500,000 eggs annually that float in the water column. Eggs hatch in 18-25 hours depending on water temperature. Salinities of water for broodfish should be around 30 ppt.

Most common production systems: Ponds stocked at rate of 2,000-4,000 per acre.

Redear Sunfish (Shellcracker)

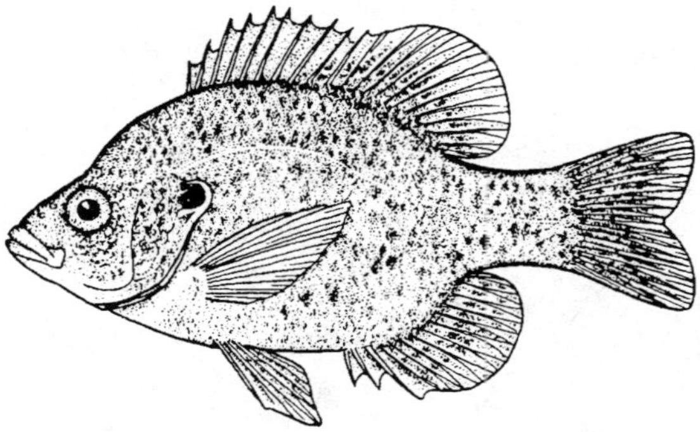

Scientific name: Lemois microlophus

Production potential: Easy

Marketing potential:

Potential: High

Size: 1-4 inches

Market: Sport

Temperature requirements:

Growing: 55-80°F

Spawning: 75-80°F Spawning starts at 75°F and is retarded at temperature below 70°F.

Lethal: NA

Feed requirements:

Protein: 32%

Fat: NA

Spawning requirements: Males construct and guard nests, usually in colonies with other males. Females may produce 15,000-30,000 eggs per spawn.

Most common production systems: Ponds

Red Swamp Crayfish

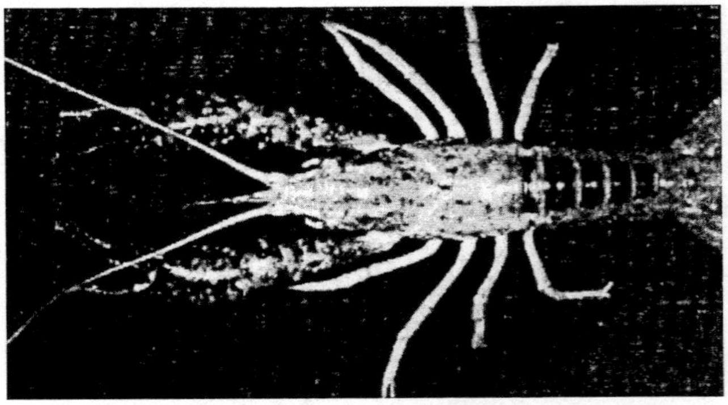

Photo by Jimmy Avery

Scientific name: Procambarus clarkii

Production potential: Easy

Marketing potential:

Potential: High

Size: Large: 15 per lb. or less

Medium: 16-20 per lb.

Small: 21 per lb or more

Market: Food and bait

Temperature requirements:

Growing: 75-85°F

Spawning: 65-75°F

Lethal: Greater than 95°F

Feed requirements:

Protein: Crayfish feed on rice stubble and zooplankton in ponds. Catfish feed (32% protein) can also be fed to crayfish.

Fat: NA

Spawning requirements: The peak of the breeding season usually occurs in May and June but may occur at other times of the year depending on temperature and other water conditions. The sperm is viable for at least eight months. Red swamp crawfish eggs hatch in 14-21 days, depending on temperature. Crayfish of the desired species are present in the area; 40-60 pounds are stocked per surface acre of water.

Most common production systems: Ponds. Crayfish are grown in combination with rice.

Smallmouth Bass

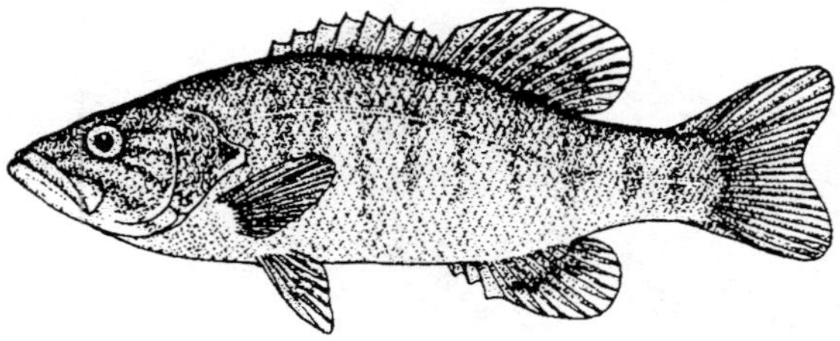

Scientific name: Micropterus dolomieui

Production potential: Easy for fingerlings. Difficult for food size.

Marketing potential:

Potential: Easy

Size: 4-6 inches for fingerlings, 1.5-3.0 pounds for food size.

Market: Sport

Temperature requirements:

Growing: 50-70°F

Spawning: 58-62°F

Lethal: NA

Feed requirements:

Protein: 45%

Fat: 8%

Spawning requirements: Nest layer, 8,000 eggs/lb body weight

Most common production systems: Ponds

Tilapia

Photo by Kevin Fitzsimmons

Scientific name: Oreochromis niloticus, O. aurea, O. mossambicus, O. hornorum

Production potential: Easy except tilapia grow slowly at temperatures lower than 70°F and die when temperatures drop into the 50°F range.

Marketing potential:

Potential: High

Size: 1.25-2.0 lb

Market: Food

Temperature requirements:

Growing: 80-87°F

Spawning: Greater than 72°F

Lethal: 55°F

Feed requirements:

Protein: 25-30%

Fat: 6-8%

Spawning requirements: Maternal mouthbrooders, spawn twice a month, 2,500 eggs/lb body weight, eggs hatch in 5-7 days. All male hybrids can be produced by crossing female O. niloticus and male O. aurea. Or by crossing female O. niloticus and male O. hornorum. Stocking ratios for fingerling production is three females to each male.

Most common production systems: Food fish stock ponds at 1500-3,000 fish per acre. Food fish production in cages stock 6-10 fish per cubic food. Stocking densities in recycle systems depend on the quality of the filtration used. However, at harvest a goal of 0.5-1.5 lb pounds per gallon is desired.

Walleye

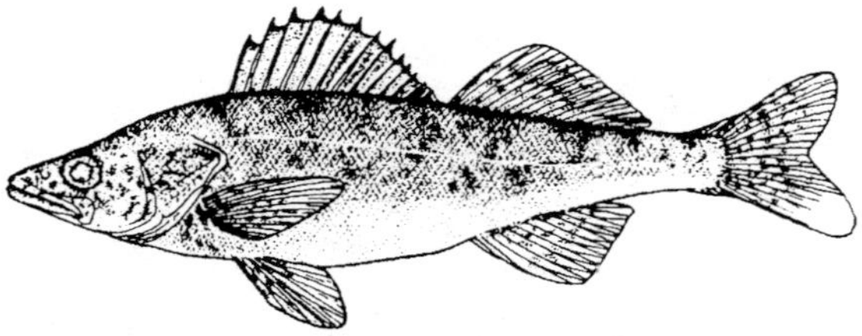

Scientific name: Stitzostedion vitreum

Production potential: Moderate for fingerlings, difficult for food size due to their cannibalistic nature

Marketing potential:

Potential: High

Size: 2-4" for fingerlings; 1-3 pounds for food size

Market: Sport and food

Temperature requirements:

Growing: 69-72°F

Spawning: 40-50°F

Lethal: NA

Feed requirements:

Protein: 45%

Fat: 10-15%

Spawning requirements: Broodfish produce 30,000 eggs/lb body weight. Spawns may be obtained by stripping eggs from ripe females and mixing with milt from males in a dry container. Mix thoroughly and then water is added to activate the sperm and permit fertilization. Eggs hatch in around 10 days at 44-55°F.

Most common production systems: Fingerlings are produced in fertilized ponds stocked at 100,000-150,000 fry per acre. Food fish may be produced in ponds or cages if cannibalism is prevented by grading to maintain a uniform size.

Yellow Perch, Lake Perch

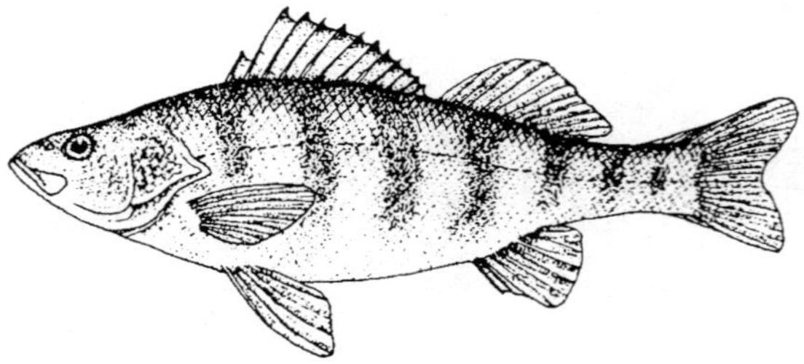

Scientific name: Perca flavescens

Production potential: Moderate, due to slow growth of males

Marketing potential:

Potential: High

Size: 3-5 fish per pound

Market: Food

Temperature requirements:

Growing: 55-78°F

Spawning: 50°F

Lethal: NA

Feed requirements:

Protein: 32-40%

Fat: NA

Spawning requirements: Fish spawn in the spring when water temperatures reach 50°F. Eggs are contained within a gelatinous ribbon. A 0.5 lb female will produce 10,000-20,000 eggs. Eggs hatch in approximately two weeks at 53°F. Spawns may be obtained by stripping eggs from ripe females and mixing with milt from males in a dry container. Mix thoroughly. Water is added to activate the sperm and permit fertilization.

Most common production systems: Ponds stocked at 200,000-300,000 fry per acre for production of 3-5 inch fingerlings. Fingerlings are stocked at 2,000 per acre for production of food size fish.

White Sucker

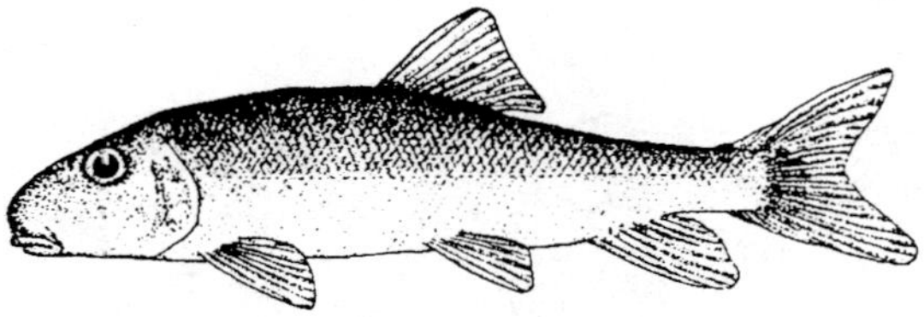

Scientific name: Catostomus commersoni

Production potential: Easy

Marketing potential:

Potential: High

Size: 6-10 inch size are used for northern pike, muskellunge, and large walleye. Smaller suckers can be used for other species.

Market: Bait

Temperature requirements:

Growing: 65-75°F

Spawning: 49-59°F

Lethal: 90°F is the upper lethal

Feed requirements:

Protein: 32%

Fat: NA

Spawning requirements: Male congregate over spawning areas, pressing against the sides of females. Spawning is most active at dusk and dawn. Adult females produce around 30,000 eggs and randomly scatter them over the substrate. Sucker runs occur between 45-59°F. Eggs will hatch in 4-6 days at 65°F and 10-15 days at 50-60°F, but will not hatch lower than 50°F. Eggs hatched at 50-60°F are healthier and hardier than other temperatures.

Most common production systems: Ponds

Appendix

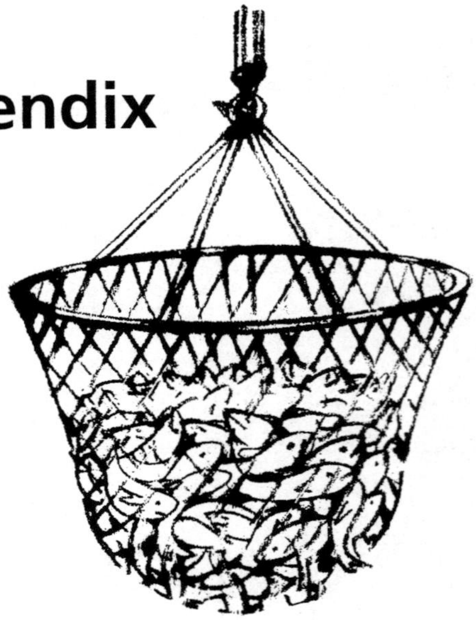

Glossary of Terms

A

Acclimate
to allow fish to slowly become accustomed to new surroundings, especially changes in temperature or water chemistry.

Acid
a substance that releases hydrogen ions.

Acre-foot
a volume measurement of water equal to one surface acre of water that is one foot deep; 325,850 gallons of water.

Acute
sudden and/or severe; usually referring to disease or stress.

Aeration
process where oxygen is added to water.

Algae
a group of plants found in water, usually microscopic, that conduct photosynthesis and produce most of the oxygen that is naturally occurring in ponds.

Alkalinity
the sum total of all bases in water; the buffering capacity of water.

Ammonia
compound containing nitrogen that is toxic to fish; the chief waste product of fish; NH_3 (unionized) or NH_4 (ionized).

Antibiotic
drug that either controls or kills pathogenic microorganisms.

Aquaculture
the husbandry of aquatic animals.

Aquatic
living in or on water for some or all of a life cycle.

Assets
the property, resources and funds owned by a business.

Automatic feeder
a device that provides feed to production units on a scheduled or metered basis.

B

Base
a substance that releases hydroxyl ions and can neutralize hydrogen ions.

Benthic
of or pertaining to the bottom of a pond or production unit.

Benthos
the community of benthic organisms.

Bloom
the green color in the water due to presence of phytoplankton algae.

Blower feeder
a mechanical feeder that uses forced air to blow the feed across the surface of production units.

Broadcast
to scatter feed or other material across a large area of a pond or production unit.

Brood stock
adult fish that are used to produce offspring.

Budget
a financial plan for allocation of assets.

Buffer
any material in water that serves to stabilize and reduce fluctuations in pH.

C

Cages
floating or suspended enclosures made of some type of net-mesh in which fish are grown or held.

Capital
money or assets obtained by selling or by borrowing to fund a potential fish farm.

Carbon dioxide
CO_2, the by-product of respiration processes; also serves as a nutrient source for plants used in photosynthesis.

Carnivorous
eating only other animals.

Cash flow
the record of actual cash transactions, both income and expense.

Chronic
long-lasting or recurring; usually referring to disease or stress.

Clean cropping; single batch
harvesting all organisms at the same time from a given production unit.

Community
the sum total of all animal or plant populations in a given environment.

Complete feed
feed that supplies 100% of the dietary nutrients required by fish.

D

Dam
an earthen embankment used to impound water; oriented perpendicular to the axis of flow of water, usually in a valley.

Demand feeder
a device that is triggered by fish movement or contact to deliver feed on demand.

Density
the number or poundage of animals in a given unit volume or area, ie., lbs/acre, lbs/cubic foot, fish/gallon, etc.

Diet
the food that is prepared and delivered to fish or other animals in a production unit.

Dissolved Oxygen
elemental oxygen (O) that is dissolved in water (DO).

Domestic
tame or bred, raised or adapted to captivity.

Drainage
the means of draining or the fate of waters leaving production units; also may refer to the movement of water that ultimately finds it way into the production unit.

E

Effluent
the water discharged from a production unit.

Embankment pond
a pond created by constructing an embankment or dam in a naturally occurring basin that collects rainfall runoff from the watershed.

Enterprise
a business venture that requires investment, management and risk in order to make a profit.

Excavated pond
a pond created by digging a hole, usually on flat or gently rolling terrain, but sometimes in land with high slopes. The dirt removed from the hole is used to build exterior levees if needed. The bottom of an excavated pond is usually below the level of surrounding terrain.

Exotic
not native to a particular area or region; especially used to refer to species not native to the USA.

Extensive culture
culture of fish at low densities, usually not requiring formulated feeds.

Extruded
having undergone a process, usually heat, causing expansion and creation of air pockets in feed, resulting in a feed that floats.

F

Fee-fishing
commercial operation dedicated to charging either access fees or poundage fees to anglers who use the facility.

Feed Conversion Ratio (FCR)
the number of pounds of feed required to produce 1 pound of fish.

Fingerling
juvenile fish; defined at various lengths for different species.

Fixed costs
costs that usually do not fluctuate due to changes in production.

Flow-through system
a system that utilizes continuous or periodic flow of water that remains in the system only a short period.

Food size
denotes fish that have reached the size that is marketable for human consumption.

Food chain
the process or complex of relationships among organisms whereby energy is transferred from producers to consumers.

Formulated feed
a manufactured feed that contains specific levels of nutrients and energy sources to meet the needs of the species for which it is produced.

Freeboard
the distance between the water surface in a pond and the top of the levee or dam.

Fry
early life stage of fish immediately following hatching.

G

Gills
the fleshy structures used by fish and some other aquatic organisms to extract oxygen from water and to release certain metabolic waste products.

Grow-out
to grow cultured organisms from juveniles to adults.

H

Hardness
the total concentration of calcium and magnesium expressed as ppm equivalent calcium carbonate.

Harvest
to remove organisms from production units.

Hatchery
facility where eggs mature and hatch into the early life stage of the cultured species.

Herbivorous
eating only plant material.

Host
organism on which a parasite or other pathogen lives.

Hybrid
a fish or plant resulting from crossing two distinctly different species.

Hybrid vigor
the characteristic of some hybrids where they outperform (growth or other trait) either of the parent species that were crossed to produce the hybrid.

I

Impound
to collect water for use in a pond or production unit.

Impoundment pond
pond created by building a dam across a land slope or basin.

Incubation
the process of protecting and nurturing eggs as they develop and progress toward hatching.

Infectious
capable of being transferred from one organism to another.

Intensive culture
culture of fish or other organisms at densities higher than can be supported in natural systems; typically requires provision of formulated feeds.

L

Label
the descriptive and safety information that is required by federal law to be attached to every pesticide that receives EPA approval and every drug that is approved by FDA.

Levee pond
a pond built on level land by pushing dirt from the center of the pond area to form exterior levees, usually on all four sides. The bottom of a levee pond is usually at or near the level of the surrounding terrain.

Levee
earthen dam or dike used to create ponds.

Lime
compound containing calcium that is used to increase the pH of soils; also increases alkalinity of water and serves to neutralize acids.

Live hauling
the transportation of live animals to a destination where they are usually restocked into other facilities.

M

Market
the buyer or sales point for a product.

Metabolism
the rate of physiologic activity in an organism.

Milligram per liter
mg/L; concentration of 1 milligram in 1 liter of water; equivalent to 1 ppm.

Mortality
death; the rate of losses due to death in a given population.

N

Nitrification
the process whereby ammonia is oxidized to form nitrite (NO_2) and then nitrate (NO_3).

Nitrite
NO_2; a compound that is toxic to many species of fish, inducing methemoglobinemia, or brown blood disease.

Non-infectious
not capable of being transferred from one organism to another.

O

Off-flavor
an objectionable flavor in fish.

Omnivorous
eating anything.

Osmoregulation
the process whereby animals maintain appropriate levels of dissolved minerals and gases in their bodies through the process of osmosis.

Ovaries
the paired female reproductive organs in fish.

P

Parasite
an organism that lives in or on another organism, usually causing harm or stress to the host organism.

Parts per million
a concentration of any material in water such that 1 pound of the material is dissolved in 1 million pounds of water; equivalent to mg/l.

Pathogen
a disease causing organism.

Pathogenic
disease-causing.

Pesticide
any compound or material that is used to control pests; usually categorized as either herbicide, insecticide or fungicide.

pH
an index to the acidity of a compound; the negative logarithm of the hydrogen ion concentration.

Photoperiod
measurement of length of exposure to light (either sunlight or artificial light).

Photosynthesis
the process whereby plants utilize energy from the sun, along with carbon dioxide and water, to produce food for themselves and to create a by-product surplus of oxygen.

Phytoplankton
microscopic plants.

Plankton
microscopic plants and animals.

Polyculture
production system where more than one species or kind of organism is grown.

Population
group of fish in a given production unit; usually refers to fish of the same species.

Processing plant
a facility where fish or other products are prepared for distribution to food markets, usually taking the product from live form to butchered and packaged forms.

Prolific
producing high numbers of offspring.

Propagation
raising or breeding of some organism.

Prophylactic
preventing or protecting from disease.

R

Raceways
narrow, rectangular units that allow a frequent water exchange by water flowing through the units, usually by force of gravity, but sometimes through a pump system. The ratio of depth:width:length is often 1:3:30 (for example, a raceway 3 feet deep might be 9 feet wide and 90 feet long). Raceways are frequently used for rearing trout, and lend themselves to use in areas with steep slopes, but can be used in any topography.

Recirculating system
a system that receives little or no fresh water, but reuses water by movement through a filter and then back into the system.

Retail
sale of fish or product to consumers.

Ripe
referring to fish, indicates ready to spawn.

Runoff
water that flows across the earth's surface following rain events; often captured to fill production ponds.

S

Sac fry
fry that have not absorbed the yolk sac that is attached to the abdomen after hatching from the egg.

Salinity
the concentration of all ionic salts in water.

Secchi disk
a disk used to measure the turbidity of water.

Sediment
materials such as soil that settle out of the water column to the bottom of a production unit.

Seed Stock
juveniles of species that are to be grown in production units to market size.

Seepage
the loss of water through the bottom of a pond or other production unit.

Seine
a net used to harvest fish from ponds; usually 5-10 feet high and up to several hundred feet long, with floats at the top and weights at the bottom.

Spawning
the management of adults to produce eggs.

Species
the most definite classification of living organisms such that all organisms within a species share specific characteristics and are capable of breeding to produce more individuals of the same species.

Standing crop
the total weight of fish in a production unit.

Stratification
the layering effect in water due to temperature differences.

Stress
the physical effect of environment that places a fish or other organism at a disadvantage; the inability to handle or cope with changes in the environment.

Stressor
any factor that induces stress in an animal or plant.

Substrate
soils or other materials in the bottom of the production unit.

Supplemental feeds
feeds that are use to provide nutrition in addition to natural foods in a production unit; may or may not provide all necessary nutrients.

T

TAN
total ammonia nitrogen, representing the sum total of ionized and unionized ammonia in solution in water.

Temper
to slowly change water temperature to allow fish to acclimate.

Testes
the paired male reproductive organs in fish.

Topography
the surface characteristics of the earth in a given region or location; often referred to as the "lay of the land."

Topping
partial harvest of only the market size individuals in a mixed size population.

Total length
the length of fish measured from tip of snout to end of tail.

Toxic
poisonous or harmful.

Toxicity
the relative degree to which a material is harmful or poisonous to specific organisms.

Turbidity
indication of how dark water is; an index to light penetration; usually due either to plankton or to suspended soils.

Turnover
the result of a destratification in lakes and ponds, resulting in mixing of the formerly thermally stratified layers.

V

Variable costs
costs that change due to fluctuations in production.

Virulent
highly infectious or damaging.

W

Water table
the level below the earth's surface where the soil is saturated with water; the level to which wells must be dug to provide water to the surface.

Wetland
area defined by law that is either covered with water or contains soils and plants characteristic of saturated soils; wetlands are protected by federal law.

Wholesale
sale of product to buyers who then resell the product to consumers at retail prices.

Withdrawal time
the period of time that must elapse after treatment with a drug or pesticide before an organism can be legally marketed for human consumption.

Z

Zooplankton
microscopic animals.

Index

A

B

C

D

E

F

G

H

I

L

M

N

O

P

Q

R

S

T

W

Y

SAFETY

live with it

TAKING THE GUESSWORK OUT OF AQUACULTURE

Aquaculture is the fastest growing segment of agriculture, with about 100 different species of fish and shellfish cultivated on U.S. farms. *Taking the Guesswork Out of Aquaculture* is a basic guide for prospective producers of aquaculture products to use in evaluating their potential for success. Chapters are devoted to the following topics:

- **Types of aquaculture enterprises**
- **Commercial catfish production**
- **Required resources and investment**
- **Site selection**
- **Stocking**
- **Feeding**
- **Water quality**
- **Fish health and disease management**
- **Harvesting**
- **Marketing**

Also included in *Taking the Guesswork Out of Aquaculture* are a decision checklist, common calculations used in fish pond and lake management, species profiles, and a section on additional sources of aquaculture information.

A glossary of aquaculture terms is included.

JOHN DEERE PUBLISHING
The all-encompassing source

Deere & Company

FBM 20101NC
ISBN 0-86691-252-5
Litho in U.S.A.